Lecture Notes in Electrical Engineering

Volume 846

The book series *Lecture Notes in Electrical Engineering* (LNEE) publishes the latest developments in Electrical Engineering - quickly, informally and in high quality. While original research reported in proceedings and monographs has traditionally formed the core of LNEE, we also encourage authors to submit books devoted to supporting student education and professional training in the various fields and applications areas of electrical engineering. The series cover classical and emerging topics concerning:

- Communication Engineering, Information Theory and Networks
- Electronics Engineering and Microelectronics
- Signal, Image and Speech Processing
- Wireless and Mobile Communication
- Circuits and Systems
- Energy Systems, Power Electronics and Electrical Machines
- Electro-optical Engineering
- Instrumentation Engineering
- Avionics Engineering
- Control Systems
- Internet-of-Things and Cybersecurity
- Biomedical Devices, MEMS and NEMS

For general information about this book series, comments or suggestions, please contact leontina.dicecco@springer.com.

To submit a proposal or request further information, please contact the Publishing Editor in your country:

China

Jasmine Dou, Editor (jasmine.dou@springer.com)

India, Japan, Rest of Asia

Swati Meherishi, Editorial Director (Swati.Meherishi@springer.com)

Southeast Asia, Australia, New Zealand

Ramesh Nath Premnath, Editor (ramesh.premnath@springernature.com)

USA, Canada:

Michael Luby, Senior Editor (michael.luby@springer.com)

All other Countries:

Leontina Di Cecco, Senior Editor (leontina.dicecco@springer.com)

**** This series is indexed by EI Compendex and Scopus databases.****

More information about this series at https://link.springer.com/bookseries/7818

Santiago Berrezueta · Karina Abad
Editors

Doctoral Symposium on Information and Communication Technologies - DSICT

Springer

Editors
Santiago Berrezueta
CEDIA
Cuenca, Ecuador

Karina Abad
CEDIA
Cuenca, Ecuador

ISSN 1876-1100 ISSN 1876-1119 (electronic)
Lecture Notes in Electrical Engineering
ISBN 978-3-030-93717-1 ISBN 978-3-030-93718-8 (eBook)
https://doi.org/10.1007/978-3-030-93718-8

This Springer imprint is published by the registered company Springer Nature Switzerland AG
The registered company address is: Gewerbestrasse 11, 6330 Cham, Switzerland

Preface

The first edition of the Doctoral Symposium on Information and Communication Technologies—DSICT 2021 took place in Guayaquil, Ecuador, from November 24 to 26, 2021, in hybrid mode (virtual/presencial) at the Universidad Politécnica Salesiana (UPS) Campus. DSICT look becomes the most important event for doctoral students in the field of ICTs in Ecuador. This scientific event brings together researchers, educators, professionals, and students from different parts of the world for the dissemination of results and research activities regarding the development, problems, and forecasts of the use of ICTs in multiple fields of application. This event is organized by Corporación Ecuatoriana para el Desarrollo de la Investigación y la Academia (CEDIA).

This Symposium Doctoral includes the presentation of several manuscripts related to ICT in different fields such as biomedical sensors and wearables systems, data science, ICT's applications, Industry 4.0, smart cities, software development, technology, and environment.

In addition to the presentation of the articles, the doctoral student shared the following:

- The summary of his/her research with the motivation justifying the importance of the research problem,
- Research questions and hypotheses formulating what you want to achieve, solve, or demonstrate,
- Technical challenges associated with the research questions and the discussion of existing solutions with an explanation of why they do not solve the problem,
- The discussion on the originality of your contributions stating how your results go beyond existing approaches,
- The description of the research methods including a strategy for evaluating your work to provide credible evidence of your results, and
- The description of the progress you made so far toward completing the research.

The first edition of DSICT received 44 manuscripts from 29 authors from six countries. All these articles passed the control of the maximum percentage of similarity with previous works, which were reviewed by peers by the DSICT program committee, which consists of 66 highly experienced researchers from 27 countries. To ensure a high-quality and careful review process, we assign at least three reviewers to each article. Based on the peer-review results, 11 full articles were accepted, representing an acceptance rate of 25%.

November 2021

Santiago Berrezueta
Karina Abad

Organization

Honorary Committee

Cecilia Paredes	CEDIA, Ecuador
Juan Cárdenas Tapia	Universidad Politécnica Salesiana, Ecuador
Juan Pablo Carvallo Vega	CEDIA, Ecuador

General Chair

Juan Pablo Salgado Guerrero	Universidad Politécnica Salesiana, Ecuador

Program Committee Chairs

Santiago Berrezueta-Guzman	CEDIA, Ecuador
Karina Abad	CEDIA, Ecuador

Program Committee

Agustin L. Herrera-May	Universidad Veracruzana, Mexico
Agustín Yagüe	Universidad Politécnica de Madrid, Spain
Alex-F. Buitrago-Hurtado	Universidad Externado de Colombia, Colombia
Alvaro Suarez	Universidad de Las Palmas de G.C., Spain
Angel Alberto Magrñan	Universidad de La Rioja, Spain
Arcangelo Castiglione	University of Salerno, Italy
Artur Rydosz	AGH University of Science and Technology, Poland
Belen Bermejo	University of the Balearic Islands, Spain
Belen Curto	Universidad de Salamanca, Spain
Carlos Abreu	Instituto Politécnico de Viana do Castelo, Portugal
Carme Quer	Universitat Politècnica de Catalunya, Spain

Natasa Zivic	University of Siegen, Germany
Noman Naseer	Pusan University, South Korea
Noor Zaman	Taylor's University, Malaysia
Patricio Galdames	Universidad del Bio-Bio, Chile
Piotr Dziurdzia	AGH University of Science and Technology, Poland
Raúl Antonio Aguilar Vera	Universidad Autónoma de Yucatán, Mexico
Robert-Alexandru Dobre	Politehnica University of Bucharest, Romania
Roberto Murphy	INAOE, Mexico
Rosaria Rucco	University of Naples Parthenope, Italy
Rostom Mabrouk	Bishop's University, Canada
Rui Zhao	University of Nebraska Omaha, USA
Ruoyu Su	Memorial University of Newfoundland, Canada
Samuel Ortega	University of Las Palmas de Gran Canaria, Spain
Sara Paiva	Oviedo University, Spain
Stavros Souravlas	University of Macedonia, Macedonia
Stefano Mariani	Università degli Studi di Modena, Italy
Tomasz Bieniek	Institute of Electron Technology, Poland
Valerio Baiocchi	La Sapienza' University, Italy
Vera Ferreira	Federal University of the Pampa, Brazil
Wojciech Zabierowski	Lodz University of Technology, Poland
Yanhua Luo	University of New South Wales, Australia
Zoltán Adam Tamus	Budapest University of Technology, Hungary

Contents

Linear Architecture Mixed-Routing Scheme Based Wireless Sensor Networks 6G Communication: From Orchestrating to Reshuffling

Amal Chaffai(✉)

Communications Department, Universitat Politècnica de València, València, Spain
amcha@doctor.upv.es

Abstract. As one effective way to be a confluence of past trends with higher level of connectivity. And to reshuffle the convergence connected things to upcoming connected intelligence things. A sustainable wireless sensor networks communication infrastructure for new generation environment. Need stringents energy efficiency, maximum network lifetime, and full coverage. Wherefore, our paper focuses mainly on enhancing a reduction energy consumption of wireless sensor networks. In order to extend the lifetime of wireless sensor networks, while ensuring less holes than possible and improving the coverage. The study consists in the network performance analysis of linear architectures mixed-routing scheme with $3hops$. As one of the applications for data compression. The distance d reduction leads to a gradual decrease in the energy consumption.

Keywords: Linear architecture · Less holes · Balancing energy · Maximizing network lifetime · Full coverage

1 Introduction

In the advent new generation era, users will expect seamless high-end services. The new platforms, have to provide seamless communication even in such harsh environments, without constraints of time and physical location [1,17]. For instance, in a $6G$ environment, artificial intelligence (AI), could be a representation of a remotely controlled set of sensors management of real world and problem detection and mitigation can be done efficiently. Nevertheless, uneven energy depletion causes energy holes and leads to degrade network performance [18]. According to [19], coverage issue is defined as a minimization problem, which is the total area of the coverage holes in a network need to be minimized as small as possible. For this purpose, sensors are organized in a hierarchical way using the most common practical solution, the clustering technique. Since can provide the spatial reuse of resources to increase system capacity. As well as, can share the routing information with other cluster-heads (CHs) or cluster

S. Berrezueta and K. Abad (Eds.): *Doctoral Symposium on Information and Communication Technologies - DSICT*, LNEE 846, pp. 1–11, 2022.
https://doi.org/10.1007/978-3-030-93718-8_1

gateways. In addition, this restriction reduces the number of transmissions performed for distributing routing information. Accordingly, the connectivity can be increased thanks to the reduction of the overhead. Moreover, the effective load balancing by a reduction in the size of the routing table, lead to stabilize the network topology. Wherefore the communication within a cluster, as well as between different can be organized as a combination of single - hop and multi - hop communication. For example, one clear application of multihoping is Vehicle-to-everything ($V2X$) its usefulness to empower the vehicles with the higher level of connectivity. Enabling seamless end-to-end personal mobility, to report instant data on traffic, environmental conditions and public safety. Thus, to clust ($WSNs$) we could apply some well know algorithm such as in [20]; however the development of clustering algorithms is outside the scope of this works, see [24]. In this paper, we aim to provide readers with a comprehensive overview of various aspects related to linear architecture mixed-routing scheme, based on $3hops$ technique. The remainder of this paper is organized as follows: Sect. 2 reviews related work, Sect. 3 addresses ($WSNs$) infrastructure scenario. Sections 4 linear architecture. And Sects. 5 clusters-heads communication. Then, in Sect. 6 we show some illustrative results of energy consumption and the coverage area. The contribution ends, with a final discussion of the results obtained followed by a conclusion and future work are reported in Sect. 7.

2 Related Work

The problem of the energy gap is a key factor that affects the useful life of the network [1]. There has been a good number of research works to find an effective solution to the prevention of energy depletion near the sink of a wireless sensor network. Energy efficiency has been considered as the most important design challenge in sensor networks [10–12]. Since sensor nodes often are deployed in tactical environments, it is difficult or impossible to recharge or replace batteries (in, e.g. $M2M$) [5,13,14].

In recent years, part of the research in ($WSNs$) has been aimed at studying the potential of collaboration between the different nodes in the data collection processes and their aggregate processing data aggregation. With the main objective of minimizing the impact that the redundant transmissions have in the joint performance of the network and in the exhaustion of the battery of the nodes. Reason for which the routing designed for a ($WSNs$) is centered precisely in this sense, constructing hierarchical algorithms that optimize on the one hand the distance of the resulting route and energy consumption. And on the other hand, its duration and ability. A well-known hierarchical architecture addressing algorithm is $LEACH$ [21] (W. Heinzelman, A. Chandrakashan and H. Balakrishnan) that use randomization to rotate the cluster-heads (CHs) and achieve a significant improvement compared to a direct approximation. Therefore, it would be possible that the chosen cluster heads (CHs) be centralized in some part of the network and as a result some other nodes remain with no cluster head (CHs). Nevertheless, this approach allows only the grouping of (CHs) of a single hop and does not consider the problems of hotspots. Some researchers (J. Lian, L. Chen, K. Naik, T. Otzu, G. Agnew) [7] presented a mathematical model to characterize the problem of the energy gap in wireless sensor networks. Based on a traffic perspective, this analytical model examines the validity of several possible schemes that aim to mitigate or solve the problem of the energy gap. It is observed that in the networks of uniformly distributed sensors, data deployment and compression of data have a positive effect. (J. Lian, K. Naik and G. B. Agnew) [7] otherwise, have considered the use of competition radio to produce unequal grouping. But this mechanism does not consider the density of the deployed sensors that could cause a lot of load in the (CHs) and too much overload in the communication networks. Moreover, other researchers (V. Mhatre, C. Rosenberg) [13,14] proposed an unequal grouping mode for the organization of the network that showed an unrealistic architecture because they considered that scalability is not a concern. Therefore, in some literature available at hand, availability of the strong infrastructure for a new application as $5G$ should be configuring to pave a way to sensed area to roadmap toward vision $6G$ [8].

3 WSNS Infrastructure Scenario

As can be observed in Fig. 1, sensors are randomly distributed in a two-dimensional $(2D)$ area in both schemes. All sensors are homogeneous, equipped with identical hardware and software configuration. As well we assume omni-directional antennas with their transmission ranges as circles. We do not suppose any limitation in the range of coverage. Except the attenuation due to distance, we assume a perfect radio channel, there are no fast or slow fading effects, nor bit error [6,16,23]. There is only a single transceiver unit per node. For cluster configuration purposes, we assume that all nodes know their actual physical location in the $(2D)-(WSN)$ area by applying some known location technique. Also, the topologies has been studied assuming the maximum number of rings, $r_k = 5$.

4 Linear Architecture

The main goal of our contribution aim to provide a theoretical study with $3hops$ in a mixed routing scheme in linear architecture. In order to seek the ways to avoid the creation an energy hole around the sink. Since this work has been taken as a complimentary existing works on designing random uniform node deployment schemes [2]. Accordingly, [3,4] we assume a wireless sensor network that consists of a sink and a set of sensors randomly deployed in its broadcast range as illustrated in Fig. 1. For simplicity, once again we assume that the sink node is centrally placed, although this is not really necessary. Hence, the coordinates has been introduced in cartesian-like form $(x, y)c$ in the following referred to delineate area as cluster cell coordinates such that the sink is located at the center delineate area as cluster of cell $(0, 0)c$. For later convenience, as well we introduce coordinates in polar (or ring) form $(u, v)p$. We call these coordinates

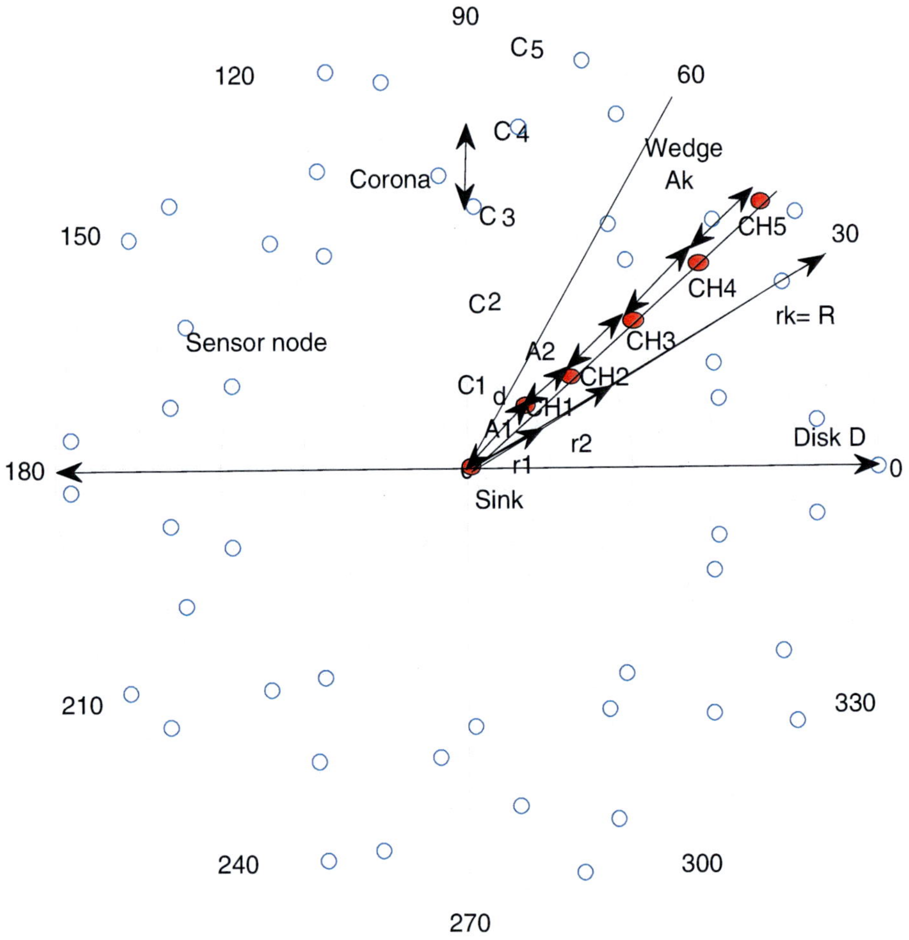

Fig. 1. A wedge W partitioned into k sectors in linear architecture

polar coordinates in the following. Additionally, we introduce the notion of axes (± 30) degrees originating at sink and sectors [3,16] (Table 1).

5 Linear Approach

A specific wedge used to show how can be affected in a working system. And can cover a plane with no gaps or overlaps. The sensor network model consider for example, an area the disk D of radius R partitioned into five coronas C_1, C_2, C_3, C_4 and C_5 as illustrated in the Fig. 1. Then, also an arbitrary wedge subtended by an angle of θ. W(wedge) is partitioned into K sectors as $A_1, A_2, \ldots,$ A_K, by its intersection with K concentric circles, centered at the sink, and of

Table 1. Table of sectors Sk and axes Ak

Axis	Condition	Sector	Condition
A_0	$0 = y_h < x_h$	S_0	$0 < y_h < x_h$
A_1	$0 < y_h = x_h$	S_1	$0 < x_h < y_h$
A_2	$0 = x_h < y_h$	S_2	$x_h < 0 < y_h$
A_3	$x_h < y_h = 0$	S_3	$x_h < y_h < 0$
A_4	$x_h = y_h < 0$	S_4	$y_h < x_h < 0$
A_5	$y_h < x_h = 0$	S_5	$y_h < 0 < x_h$

monotonically increasing radii $R = r_k$ [23]. Then, we can implement the areas of clustering sectors into following a probability transition equations:

$$S_i = \frac{\Theta}{2} * (r_i^2 - r_{i-1}^2) \tag{1}$$

$$A_i = S_i + A_{i-1} \; ; \; 1 \leqq i \leqq 3. \tag{2}$$

Once sensory data is collected by (CHs), in order to minimize the amount of traffic to the sink node. We assume that the cluster identified by (i, j) that is, the set of sensor nodes located in sector $A_{i,j}$, where i is the corona identifier, and j is the wedge identifier, are to perform a certain task T. A number of sensors in sectors $A_{1,j}, A_{2,j}, \ldots, A_{i-1,j}$ are selected to act as routers of the data collected by the sensors in $A_{i,j}$ to the sink [3,4].

6 Clusters-Heads Communication

6.1 Cluster Size

Since in all the systems clustering have the same concept as aforementioned in [5]. We can use for any two positive integers or shift parameters (i, j) and $j \leqslant i$ the number of clusters operating at different frequency is given by $N = i^2 + ij + j^2$. The communication range required for both types of communication will have dependent of the number of rings, rk that conform the wireless sensors networks $(WSNs)$.

6.2 Linear Architecture Communication

In order to minimize the total amount of energy spent on routing along a path originating at a sensor in corona A_i and ending at the sink, all the coronas must have the same width d and the optimal amount of energy is i times the energy needed to send the desired information between adjacent coronas. Thus, we should set optimal deterministic value r_1 and determine $r_2, r_3, \ldots, r_k$. In a generic sector A_i it follows that the transmission range needed to send information between A_i and A_{i-1} is $r_i - r_{i-1}$ [23]. Being that, the width of each corona is $\Delta_i = r_i - r_{i-1}$. And each sensor has a maximum transmission range, denoted by R. The widths of the coronas must satisfy the following equality [23], in order to tailor the coronas and the energy expenditure be balanced across them: $r_1 = r_1 - r_0 = r_2 - r_1 = r_3 - r_2 = r_4 - r_3 = r_5 - r_4 = \ldots = r_i - r_{i-1}$, since $r_0 = 0$.

6.3 Hoping Techniques

To help relieve the burden of long distance transmission for the nodes far away from the sink. Each node, in direct transmission mode, sends its data directly to the sink without any relay [15]. In hop-by-hop transmission mode, each node forwards the data to its next hop neighbors.

6.4 Load Balancing Inter-rings

Since the fundamental assumption adopt an hybrid - routing [16,22]. On sector A_i each event in (CH) node is propagated to A_{i-1} with probability p_i, while with probability $1 - p_i$ it is propagated directly to the sink. In other words, we want to transfer this traffic load to the (CH) in sector A_{i-1}. Then, (CH) in sector A_{i-1} has to cope with $1 + 1/A_{i-1}$ times the per (CH) load of sector A_i plus the load originating from A_{i-1} itself, as illustrated in Figs. 3, 4. $(x, x) \longrightarrow (x-1, x-1)$. In addition, non-axes (CHs) transfer the information to two inner-ring neighbors, for example $(x-1, x) \longrightarrow (x-1, x-1)$ with probability $p_{(x-1,x)-(x-1,x-1)}$, $p_{(x-1,x)-(x-1,x-1)} + p_{(x-1,x)-(x-2,x-1)} = q$. From this we can assume once again that $1 = p_{x-1} + p_x$, hence $p_x = 1 - p_{x-1}$, where $P_{x-1} = q$. With simple manipulations we can find the value of $r_k = r_1$ minimum communication range required. The routing probabilities are chosen according to the load balancing principle between (CHs) of the same ring. For instance, let us consider the routing probabilities from ring x to ring $x-1$, where the traffic load in each (CH) is e_k, with $k = x, x-1$ respectively. Apart from the local traffic, the load received at $(CH)(x-1, x-1)$, e_{x-1}, is equal to $e_{x-1} = p_{(x-1,x)} - (x-1,x)e_x + e_x + p_{(x,x-1)-(x-1,x)}e_x = (2p_{(x-1,x)-(x-1,x-1)} + 1)e_x$ where the last equality follows from the symmetry of the load balancing, i.e., $p_{(x-1,x)-(x-1,x-1)} = p_{(x,x-1)-(x-1,x)}$. In addition, the load received at $CH(x-2, x-1)$, also e_{x-1}, is equal to $e_{x-1} = p_{(x-2,x)-(x-2,x-1)}e_x + p_{(x-1,x)-(x-2,x-1)}e_x = 2p_{(x-2,x)-(x-2,x-1)}e_x$, since $p_{(x-2,x)-(x-2,x-1)} = p_{(x-1,x)-(x-2,x-1)}$. The generalization to arbitrary rings and sectors can be found in table I of [16].

7 Energy Consumption and Coverage Area

A fundamental assumption here is, we adopt the following general energy consumption model [15].

$$Et(d) = ad^{\alpha} + b \text{ and } Er = b, \tag{3}$$

By normalizing

$$Et(d) = d^{\alpha} + c \tag{4}$$

where, $a > 0$ is a constant standing for the transmitter amplifier, $b > 0$ is a constant representing energy for running electronic circuit, where c is a technology-dependent positive constant and the path loss α is

$$2 \leqq \alpha \leqq 6. \tag{5}$$

Then each case has been analyzed in a specific way according to its suitable parameters to have an efficient energy expenditure. The main goal of this section is to prove the above assumptions. To evaluate the performance of the proposed algorithm with the uniform node distribution. Using the simplified energy model stated in Sect. 4, without consider the MAC layer and physical layer issues.

The generic computation for a case $\alpha > 2$ apply the equation follow (the derivative of this equation can be find in) [23]:

$$E_i = \frac{T}{\pi\rho}[1 - \frac{r_i^2 - 1}{r_k^2}]\frac{(r_i - r_{i-1})^{\alpha} + c}{r_i^2 - r_{i-1}^2}; i = 1, 2, 3, \ldots \tag{6}$$

Hence, we can illustrate $E_1, E_2, \ldots, E_k$ by a numerical example [5,23]. And the system parameters are:

Algorithm: Expenditure Energy
1- Inicialize the parameters T, r_k, c
2- Conversion coordinate
3- Camputing r_1, r_2, r_3, ..., r_k
4- Computing energies E_1, E_2, ..., E_k
5- Output the results of energies E_1, E_2, ..., E_k

System Parameters:	
Radius	$R = 225\,\text{m}$
A positive constant	$c = 4500$
Path loss	$\alpha = 4$
Density (nodes/cluster)	$\rho = 35$

Let T equals the total number of tasks that the wedge can handle during the lifetime of the network (Fig. 2).

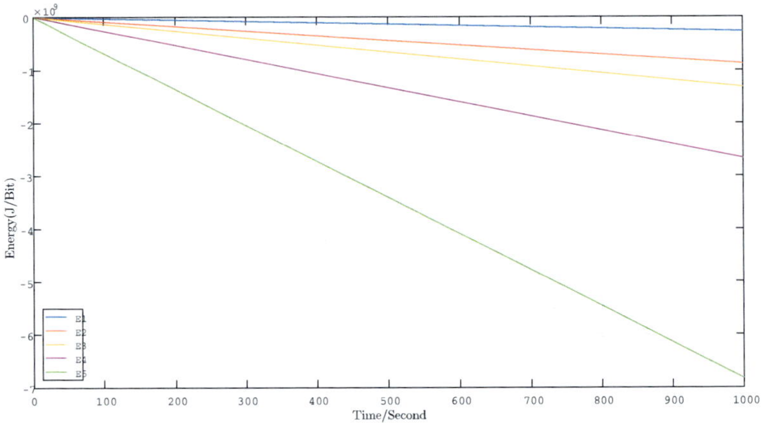

Fig. 2. Expenditure energy with 2 hops

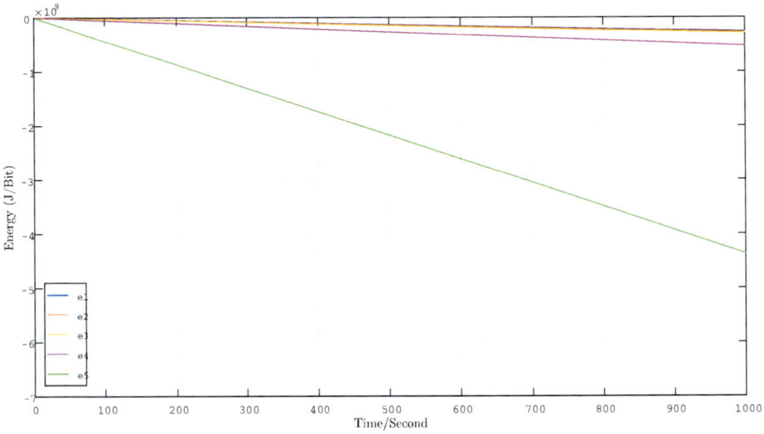

Fig. 3. Expenditure energy with 3 hops

Then, the energy consumption per (CH) significantly low has been ascertained. As synopsis of energy expenditure in a linear architecture [3,4,23], to minimize the total energy spent per routing path. Once again, $r_0 = 0$ with A_0 the sink itself has been assumed. In the last corona, the energy expended by a node (CH) is less than 10% of the energy expended by a node (CH) in the first corona. As shown in the Fig. 4.

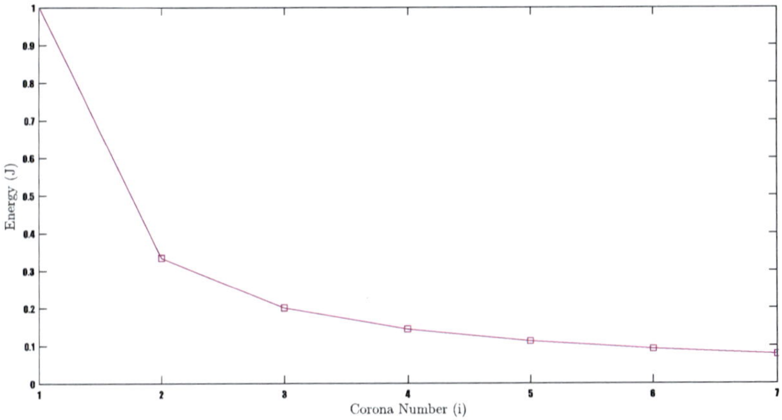

Fig. 4. Optimizing expenditure energy per path

8 Conclusion

In this contribution a theoretical infrastructure study for linear architecture wireless sensor networks ($WSNs$) has addressed the problem of energy depletion near the sink in a two-dimensional plane. The method presented has been investigated with regard to strategic locations of (CH), That is, the radius $R = r_k$ minimum communication range for each sink selected in such way the energy of sensors within this radius can be balanced with coherent effects for energy efficient communication. The monitored area has been divided into wedges comparable to the approach presented in [23]. And the assumptions mentioned before motivated by the uniformity of the deployment and each (CH) is equally likely to be the source of a path to the sink.

Therefore, the novel theoretical technique presented above with a optimized results. Show that this is the effective method to settle problems aspects of the uneven energy depletion phenomenon of balancing energy consumption and maximizing network lifetime. In addition, can serve as guideline to determine the best mode of communication in severals hostile environments. Still, for economic reasons, Bell Laboratories system designers adopted nonlinear architecture severals years ago [6,16,22]. The study will naturally be updated as we proceed with our research to overcome the shortcoming technologies [8]. Envision new generation 6G with the emerging trends, and with more stringent requirements as very high energy efficiency.

Abbreviations

The following abbreviations are used in this manuscript:

- (WSNs) Wireless Sensor Networks.
- (CHs) Cluster - heads.
- (AI) Artificial intelligence.
- (V2X) Vehicle-to-everything.

- (M2M) Machine - to - Machine.
- (x, y)c Cartesian coordinates.
- (u, v)p Polar coordinates.
- rk The minimum communication range.
- N Number of clusters.
- Ei Expenditure Energy.
- D Disk.
- R Radius.
- W Wedge.
- d distance.

Acknowledgments. This work is supported by the national project PGC2018-094151-B-I00. I thank the scientific advice of my Professor.

Conflicts of Interest. The author declare no conflict of interest.

References

1. Akyildiz, I.F., Nie, S., Lin, S., Chandrasekaran, M.: 5G roadmap: 10 key enabling technologies. Comput. Netw. **106**(4), 17–48 (2016)
2. Sari, A.: Two-tier hierarchical cluster based topology in wireless sensor networks for contention based protocol suite. Int. J. Commun. Netw. Syst. Sci **8**, 29–42 (2015)
3. Wadaa, A., Olariu, S., Wilson, L., Jones, K., Eltoweissy, M.: On training a sensor networks. Univ. York **17**, 1–8 (2005)
4. Olariu, S., Wadaa, A., Wilson, L., Eltoweissy, M.: Wireless sensor networks: leveraging the virtual infrastructure. IEEE Netw. **18**(4), 51–56 (2004)
5. Heinzelman, W., Chandrakashan, A., Balakrishnan, H.: An application - specific protocol architecture for wireless micro-sensor networks. IEEE Trans. Wireless Commun. **1**(4), 660–670 (2002)
6. Donald, V., Znati, T.: A mobility based framework for adaptive clustering in wireless ad hoc networks. IEEE J. Sel. Areas Commun. **17**(8), 66–86 (1999)
7. Lian, J., Naik, K., Agnew, G.B.: Modeling and enhancing data capacity in wireless sensor networks. IEEE Monograph Sensor Netw. Oper. **1**, 660–670 (2004)
8. Letaief, K., Chen, W., Shi, Y., Zhang, J., Zhang, Y.A.: The roadmap to 6G: AI empowered wireless networks. IEEE Commun. Mag. **57**, 84–90 (2019)
9. de Brito, M.P.L., Peralta, M.: An analysis of localization problems and solutions in wireless sensor networks. Revista de Estudos Politcnicos **6**, 1–27 (2008)
10. Akyildiz, F., Su, W., Sankarasubramanian, Y., Cayirci, E.: Wireless sensor networks: a survey. Comput. Netw. **38**(4), 393–422 (2002)
11. Culler, D., Estrin, D., Srivastava, M.: Overview of sensor networks. IEEE Comput. **37**(8), 41–49 (2004)
12. Martinez, K., Hart, J.K., Ong, R.: Sensor network applications. IEEE Comput. **37**, 50–56 (2004)
13. Mhatre, V., Rosenberg, C.: Design guidelines for wireless sensor networks: communication, clustering and aggregation. Ad Hoc Netw. **2**, 45–63 (2004)
14. Mhatre, V., Rosenberg, C., Kofman, D., Mazumdar, R., Shroff, N.: A minimum cost heterogeneous sensor network with a lifetime constraint. IEEE Trans. Mob. Comput. **4**(1), 4–15 (2005)

15. Rodoplu, V., Meng, T.: Minimum energy mobile wireless networks. In: IEEE International Conference on Communications, vol. 3, pp. 33–39 (1998)
16. Casares, V., Wuchner, P., Paramo, D.P.: Combined contention and TDMA - based communication in wireless sensor networks. In: 8th Euro - NF Conference on Next Generation Internet (NGI), Karlskrona, Sweden, pp. 1–8 (2012)
17. Bhardwaj, M., Garnett, T., Chandrakasan, A.P.: Upper bounds on the lifetime of sensor networks. In: IEEE International Conference on Communications. Conference Record 2001 (Cat. No. 01CH37240), vol. 3, pp. 785–790 (2001)
18. Tullberg, H., et al.: The metis 5G system concept: meeting the 5G requirements. In: Proceedings of the 2000 IEEE Wireless Communications and Networking Conference, Chicago, IL, USA, vol. 54, pp. 132–139 (2016)
19. Shen, Z.W.X., Chen, J., Sun, Y.: Grid scan: a simple and effective approach for coverage issue in wireless sensor networks. In: IEEE International Communications Conference, pp. 80–84 (2006)
20. Gerla, T.M., Pei, G.: On demand routing in large ad hoc wireless networks with passive clustering. In: Proceedings of the 2000 IEEE Wireless Communications and Networking Conference, Chicago, IL, USA, pp. 100–105 (2003)
21. Heinzelman, W.: Application - specific protocol architectures for wireless sensor networks, Doctoral Dissertation, MIT (2000)
22. Goldsmith, A.: Wireless Communications. Cambridge University Press, Cambridge (2005)
23. Stojmenovic, I., Olariu, S.: Data-centric protocols for wireless sensor networks. In: Handbook of Sensor Networks: Algorithms and Architectures, pp. 417–456. Wiley (2005)
24. Martin, E., Liu, L., Covington, M., Pesti, P., Weber, M.: Location Based Services Handbook Applications, Technologies, and Security. Cambridge University Press, Boca Raton (2005)

Ethnography of Irrigation Technology: A Unique Case Study in the Puichig Agricultural Association, Ecuador

Cristian Mosquera-Cruz(✉)

Escuela Politécnica Nacional, Quito, Ecuador
cristian.mosquera@epn.edu.ec

Abstract. Currently, agricultural and livestock production constitutes one of the most important economic incomes worldwide, particularly in countries located in the southern hemisphere, where modernized irrigation has been implemented. In this context, a unique case study is proposed from an ethnographic perspective on the social and cultural management of irrigation in the Puichig Agricultural Association. To analyze the social and cultural consequences of implementing green revolution technology in agricultural production systems in the Machachi parish in the Mejia canton. Producing knowledge concerning the social and cultural consequences that emerge from the complex and structural relationships between irrigation technologies and agricultural production systems, especially in countries in the global south, such as Ecuador, which is developing and technology-dependent.

Keywords: Irrigation technologies · Environment · Agricultural production systems · Ethnography

1 Approach to the Subject and Object of Study

Countries are facing unprecedented challenges in their efforts to achieve sustainable development. The COVID-19 pandemic is threatening lives, livelihoods, and entire economies, erasing decades of economic progress, poverty reduction, and gains in human development [1]. Current estimates are that nearly 690 million people are hungry, or 8.9% of the world population – up by 10 million people in one year and by nearly 60 million in five years [2]. Intensifying water constraints threaten food security and nutrition. Thus, urgent action is needed to make water use in agriculture more sustainable and equitable. Irrigated agriculture remains by far the largest user of freshwater, but scarcity of freshwater is a growing problem owing to increasing demand and competition for freshwater resources [3].

In this sense, the agricultural production system is currently related to the increase - in terms of efficiency - of land use, originated by world demand for food and global indicators to reduce hunger and poverty in the face of population growth. The cost of a healthy diet exceeds the international poverty line (established at USD 1.90 purchasing power parity (PPP) per person per day), making it unaffordable for the por [2].

S. Berrezueta and K. Abad (Eds.): *Doctoral Symposium on Information and Communication Technologies - DSICT*, LNEE 846, pp. 12–20, 2022.
https://doi.org/10.1007/978-3-030-93718-8_2

The United States (US) and international development agencies such as the United Nations (UN), the Food and Agriculture Organization of the United Nations (FAO), The Organization of American States (OAS) and the International Monetary Fund (IMF), encouraged the expansion of agribusiness in the global south, developing livestock industries, supplied with US grains and generating the need to introduce green revolution technologies, in order to expand and at the same time influence the policies and social representations of the countryside. The green revolution technologies constitute a specific input package composed of hybrid and other high-yielding plant varieties, mechanization, agrochemicals, fertilizers, pesticides, and irrigation. It was first developed in the 1930s in the USA as modern agriculture and then exported to the global south countries as the "green revolution" [4].

For the Food and Agriculture Organization, the area of the world cultivated in 2013 was 348,791,922 ha, destined to obtain raw materials such as corn, sugar cane, soybean, African palm, and others, oriented worldwide as an important alternative energy source to the use of fossil fuels, thus impacting land use, prices and food availability [3]. In this perspective, Azar & Johansson analyze the possible effects of biomass production on food security and poverty reduction, the link between energy markets and agriculture, and strategic changes in the use and suitability of agricultural land [5].

In addition, the industrialization of agriculture and livestock, through green revolution technologies and other current strategies, is possibly related to rural inequalities, manifested in economic inequalities such as uneven distribution of land and water, unemployment, underemployment, or insecure employment. Additionally, lack of assets and credit; political inequalities that originate in the lack of guaranteed individual and collective rights, limited access to the local and national power system; cultural inequalities such as the lack of recognition and segregation of cultural values and practices, as well as access to information and knowledge.

Concerning the above, Friedmann & McMichael suggest that it is necessary to elucidate how power relations, especially those of food production and consumption, are fundamental for the functioning, reproduction, and hegemony of global capitalism, through the construction of new strategies and ideological discourses around the agrifood system of the global south [6].

According to the above, the configuration of agricultural production systems was transformed. Families with better socio-economic conditions were more likely to adopt the intensive package of "green revolution" technologies, which resulted in higher yields and more significant economic benefit; that is, it favored their competitive advantage over that of low-income families who had to sell their properties because they were no longer economically viable. As a result, the concentration of the best land increased, agricultural labor for poor or landless peasants decreased, and farmers were relocated to less suitable agricultural areas without adequate technology.

According to the National Institute of Statistics and Census - INEC, 19% of the national territory is for agricultural use in Ecuador [7]. According to the National Survey of Employment, Unemployment, and Underemployment – ENEMDU, 1,500,000 people are engaged in agriculture, of which only 56% have completed primary school, and 31.8% are women [8]. The Technical Secretariat for International Cooperation - SETECI mentions that 10.48% of economic resources from international cooperation went to agriculture [9].

In contrast, according to the Plan de Desarrollo y Ordenamiento Territorial del cantón Mejía – PdyOT (Land Development and Planning Plan), there is a total population of 81,335 inhabitants of which 20.3% or 16,515 inhabitants are in the urban area, and 79, 7%, or 64,820 inhabitants are in rural areas. In rural areas, 36,165 people are the employed population, and of this total, 6,824 people, or 18.87%, are directly involved in the agricultural sector, making them the main actors in the different systems [10].

The arable land in the Mejia canton is 96,937 ha (more than 50%, i.e., 59,962 ha used for livestock and dairy production, 5,420 ha agriculture, and 1,408 ha for export crops). The moorlands, which represent high biodiversity and are the source of water regulation, cover approximately 28,017 ha [10].

The main products in the canton of Mejía include potatoes, broccoli, artichoke, barley, corn, and recently cultivating flowers. Furthermore, agricultural production systems have increased technology such as technical irrigation, greenhouses, pesticides, herbicides, chemical fertilizers, and new high-yielding crops to increase production and profits significantly [10].

The Puichig Agricultural Association is located in the highlands of Ecuador, province of Pichincha, canton Mejía, parish Machachi. It is a legally constituted organization that takes advantage of its favorable climate and has very fertile land, excellent for agricultural activities. The Association is involved in the production and marketing of vegetables, vegetables, and grains. Twenty percent of the farmers do not have access to water for irrigation, which hinders their production. Instead, they use tanks and springs for irrigation, representing an additional cost, both economically and in terms of production time; the remaining 80%, 45% have implemented drip and sprinkler irrigation systems.

2 Rationale

This article shows that, it is currently necessary to research and generate knowledge on the social and cultural consequences that emerge from the complex and structural relationships between technology and agricultural production systems, especially in the context of countries in the global south, such as Ecuador, which is in the process of development and is technologically dependent.

In this regard, the Brazilian philosopher Pinto in "the concept of technology," elaborates a philosophy of technology thought from the perspective of the periphery. He proposes an Aristotelian, anthropological and evolutionary definition as an essential mediation in the dialectic between man and nature [11]. Influenced by Hegel and Marx, Pinto conceives a dialectical typology to situate technology in relation to underdevelopment. Furthermore, he emphasizes the ideologization of technology as an essence in the domination of the center over peripheral countries.

In this sense, the contribution with the generation of new knowledge is manifest in the current relevance of interdisciplinary dialogue, between the social studies of science and technology; and the social management of agricultural production systems, in order to understand and analyze more comprehensively how the dependence of global south countries on new agricultural technologies is legitimized, significantly weakening agricultural production systems, that of peasant self-support and deepening socio-economic inequality.

In this perspective, the social importance of research is based on the contribution to description and explanation of social and cultural consequences that emerge in the implementation of green revolution technology, specifically technified irrigation- in agricultural production systems, such as the proletarianization of peasants and farmers, labor precariousness, feminization of the economy, fragmentation of the community fabric, lack of participation and social organization, health problems associated with the use of chemicals in agricultural work and food consumption.

Per the above, Feenberg proposes that technology constitutes a new type of cultural system that restructures the social world [12]. Pinch & Bijker postulate that what is needed is understanding technology from within, a body of knowledge, and a social system [13]. Winner also proposes that it is not surprising to find that technology is deeply intertwined with the conditions of modern politics [14].

3 Research Background

Feenberg proposes that two dimensions of its action must be kept in mind to understand technology fully. The first, which he calls "primary instrumentalization," is responsible for explaining the constitution of technical objects and subjects. The second, which he designates as "secondary instrumentalization" explains incorporating the objects and subjects in real technical networks. Although primary instrumentalization characterizes technological development in each society, technology also includes features that evolve historically and are articulated through a second instrumentalization that contains cultural aspects [12].

Marcuse identifies technology, as a mode of production, as the totality of instruments, mechanisms, and apparatuses that characterize the machine age and is at the same time a way of organizing and perpetuating (or changing) social relations [15]. In the same sense, Escobar states that, presently, people's daily life is consciously or unconsciously related to technology; as we shall see, it is this triad-life, work, and language-which is being destabilized by new technologies [16]. Marcuse also mentions the importance of research to define technological rationality's political character [15].

Per the aforementioned, Marcuse states that instrumental rationality has become the most potent element of reason, and therefore, of that concept that can indicate with the greatest propriety, the specific character of the project of western civilization [15]. Echeverría mentions that modern rationalism, the reduction of the specificity of the human to the development of the reasoning faculty and its reduction to how it is realized in purely technical practice or instrumentalizes the world, is thus the most direct manifestation of humanism proper to capitalist modernity [17]. In this regard, Echeverría states that capitalism was born as a form or mode of reproduction of the economic life of human beings: a way of carrying out that set of their activities, which is directly and preferably dedicated to the production, circulation, and consumption of the goods produced [17].

Tenorio states that social management is the opposite of strategic management, insofar as he attempts to replace monologic, techno-bureaucratic management with a more participatory, dialogical action, in which different social participants exercise the decision-making process [18]. This planning allows us to analyze the level of participation and decision making exercised by farmers and peasants in the adoption of specific technologies, in the design of public agricultural policies, in the commercialization of agricultural inputs, especially chemical ones by commercial companies, the loss of cultural and social practices by communities and families related to peasant agriculture, and new forms of human nutrition.

In this sense, Van der Ploeg, based on Alexander Chayanov's approach, proposes that agriculture is regulated by a set of "balances," from which decisions are made that result in the configuration of the productive process, the landscape, the participants involved, the market links and other characteristics of agricultural production. These balance processes function as principles of order presented and represented by the different participants - peasants, families, and communities [19]. He proposes understanding and describing the peasant economic unit based on its relationship with the land, resources, labor, and family.

According to Dufumier, the production system is the structured set of agricultural, livestock, and non-agricultural activities established by a producer and his family to ensure the reproduction of his farm; the result of the combination of the means of production (land and capital) and the labor force available in a given socio-economic and ecological environment [20].

The green revolution technologies constitute a specific package of inputs, consisting of hybrid and other high-yielding plant varieties, mechanization, agrochemicals, fertilizers, pesticides, and irrigation. It was first developed in the 1930s in the USA as modern agriculture and then exported to the global south countries as the "green revolution." In the context of this research, we will specifically refer to irrigation, which constitutes the irrigation technologies implemented in a production system [4].

4 Objectives and Research Question

4.1 General Objective

To analyze the social and cultural consequences of implementing green revolution technology in agricultural production systems in the Machachi parish in the Mejia canton.

4.2 Specific Objectives

- To identify the agricultural production systems of the Machachi parish located in the Mejia canton.
- To characterize the agricultural production system of the Machachi parish of the Mejia canton that uses the green revolution technology.
- To describe the green revolution technology components used in the agricultural production system of the Machachi parish of the Mejía canton.
- To analyze the social and cultural consequences of implementing green revolution technology components in the agricultural production system of the Machachi parish of the Mejia canton in the last thirty years.

- To analyze the predominant social and cultural consequences of implementing green revolution technology in the agricultural production system of the Machachi parish of the Mejia canton in the current context.

4.3 Research Question

What are the social and cultural consequences of having implemented irrigation technologies in the Puichig Agricultural Association in the Machachi parish of the Mejía canton, thirty years ago to the present?

5 Methodological Strategy

From an ethnographic perspective, a unique study concerning irrigation, social and cultural management in the Puichig Agricultural Association Analysis Unit. Concerning the above, ethnography makes it possible to describe and analyze the irrigation technologies implemented by the Association; the customs, practices, beliefs, places, spaces, and ways of life (Table 1).

In relation to the above, two different irrigation technologies are proposed for the analysis and observation, considering their constituent components: technical irrigation and gravity irrigation (Table 2).

- Technified irrigation: greater complexity and a more significant number of technological components.
- Gravity irrigation: greater complexity and fewer technological components.

Table 1. Information gathering strategy.

Levels	Instances	Method
Contextual levels	Ministery of agriculture	Document analysis
Irrigation practices	Technified irrigation project	Semistructured interviews
	Puichig agricultural association	Biographic interviews
	Gravity irrigation	Participative observation
	Technified irrigation	Focus groups

The ethnographic work will take place at the Puichig Agricultural Association located in Machachi, Canton Mejía, Ecuador.

- Participant observation as the primary method of data collection.
- Progressive targeting
- Complimentary use of documents and semistructured and biographical interviews.
- Focusing, reflexivity regarding observed performance and the technological components used for irrigating.

Table 2. Data analysis strategy.

Variable	Possible values	Theories	Concepts
Type of irrigation	Gravity irrigation	Social technology	Technological rationality
		Philosophy of technology	Intentionality/Ambivalence
	Technified irrigation	Social inequality	Agricultural proletariat
		Social management	Participation/Autonomy
		Environmental sustainability	Eco-efficiency

5.1 Design Quality

A single case study from a constructivist-based ethnographic perspective with two units was decided upon. Internal validity will be assessed through the degree to which causal relationships can be established, where certain conditions are shown to lead to others. For this, the use of the following four aspects is proposed: triangulation, adjustment to a pattern, construction of explanations, and analysis of time series.

Triangulation -an instance of integration of interpretations "and not of facts"- of data constructed by primary sources such as leaders, farmers, and technicians who accompanied the irrigation process; and secondary sources such as academic documents, regulations, and statutes of the organization, ministerial agreements, and press articles. The triangulation of information gathering techniques, based on field observation, semistructured and biographical interviews applied to primary sources and documentary review. Researcher triangulation, peer and stakeholder review.

From the qualitative perspective, external validity will be through "naturalistic" generalization related to a detailed and in-depth description of the case and its context, which is transferable and of interest to the reader and his/her experience.

Ethnographic perspective, reliability is centered on reflexivity as a guarantee of research validity, above the systematization of procedures. Reflexivity is a procedure of reliability, through which the researcher constantly asks himself about his actions in the research process, specifically about fieldwork. It is based on the protagonism acquired by the researcher as a creator of interpretations and their relationships.

6 Expected Results

- Essay on the identification of agricultural production systems in the Machachi parish of the Mejía canton.
- Study with the characterization of the agricultural production system of the Machachi parish of the Mejía canton that uses green revolution technology.
- A descriptive study of the components of green revolution technology used in the agricultural production system of the Machachi parish in the Mejía canton.

- Essay on the social and cultural consequences in implementing the components of green revolution technology in the agricultural production system of the Machachi parish of the Mejía canton in the last thirty years.
- A descriptive study on the predominant social and cultural consequences in the implementation of the technology components of the green revolution in the agricultural production system of the Machachi parish of the Mejía canton in the current context.

7 Conclusions

This paper aims to present a research proposal to describe and analyze the complex and structural relationships between irrigation technologies and the production system through a unique case study from an ethnographic perspective.

Concerning the above, the ethnography of technology, as a methodological strategy, allows producing knowledge about the social and cultural consequences of implementing irrigation technologies in the Puichig Agricultural Association.

After executing the research proposal, the objective is to present the results at next year's conference as a methodological contribution to have a model of ethnography of technology.

References

1. World Bank: World Bank Annual Report 2020. World Bank, Washington, DC (2020). https://doi.org/10.1596/978-1-4648-1619-2
2. FAO, IFAD, UNICEF, WFP and WHO: The State of Food Security and Nutrition in the World 2020. Transforming food systems for affordable healthy diets. Rome, FAO (2020). https://doi.org/10.4060/ca9692en
3. FAO: The State of Food and Agriculture 2020. Overcoming water challenges in agriculture, Rome (2020). https://doi.org/10.4060/cb1447en
4. Otero, G.: Neoliberal globalism and the biotechnology revolution: economic and historical context. In: Otero, G. (ed.) Food for the Few. Neoliberal Globalism and Biotechnology in Latin America, pp. 1–30. University of Texas Press, Austin (2008)
5. Azar, C., Johansson, D.J.A.: Valuing the non-CO2 climate impacts of aviation. Climatic Change **111**, 559–579 (2012). https://doi.org/10.1007/s10584-011-0168-8
6. Friedmann, H., McMichael, P.: Agriculture and the state system: the rise and fall of national agricultures, 1870 to the present. Sociologia Ruralis **29**(2), 93–117 (1989)
7. INEC: National Statistics Program 2017–2021. National Institute of Statistics and Censuses, Annual report 2017, Quito, Ecuador (2017)
8. ENEMDU: National Survey of Employment, Unemployment, and Underemployment, Annual report 2014, Quito, Ecuador (2014)
9. SETECI: Technical Secretariat for International Cooperation, Annual report 2015, Quito, Ecuador (2015)
10. PDyOT: Plan de Desarrollo y Ordenamiento Territorial del cantón Mejía (2015)
11. Pinto, Á.: Concepto de Tecnología, vol. 1. Contrapunto, Rio de Janeiro (2008)
12. Feenberg, A.: Critical theory of technology and STS. Thesis Eleven **138**(1), 3–12 (2017). https://doi.org/10.1177/0725513616689388

13. Pinch, T.J., Bijker, W.E.: La construcción social de hechos y de artefactos: o acerca de cómo la sociología de la ciencia y la sociología de la tecnología pueden beneficiarse mutuamente. Actos, actores y artefactos: sociología de la tecnología, pp. 19–62 (2008)
14. Winner, L.: The Whale and the Reactor: A Search for Limits in an Age of High Technology, 2nd edn. University of Chicago Press, Chicago (2020)
15. Marcuse, H.: Guerra, Tecnología y Fascismo. Textos inéditos, Medellín, Universidad de Antioquía (2001)
16. Escobar, A.: Antropología y Tecnología. Signum Editores S.A., México (1997)
17. Echeverría, B.: Ilusiones de la Modernidad. UNAM/El Equilibrista, México D.F (1997)
18. Tenorio, F.G.: Uma Alternativa: Gestão Social. Editora Unijuí, Ijuí-RS-Brasil (2016)
19. Van der Ploeg, J.D.: El Campesinado y el arte de la agricultura. Un manifiesto Chayanoviano. Universidad autónoma de Zacatecas. MAPorrúa, México (2015)
20. Dufumier, M.: Les projets de développement agricole: manuel d'expertise. Karthala, Paris (1996)

Virtual Classroom Design Model and Its Relation to Student Motivation and Performance in a Moodle Learning Environment During the Emergency of COVID-19

Joe Llerena-Izquierdo[1,2](✉)

[1] Universitat Oberta de Catalunya, Barcelona, Spain
jllerenai@uoc.edu, jllerena@ups.edu.ec
[2] Universidad Politécnica Salesiana, Guayaquil, Ecuador

Abstract. Virtual learning environments during the pandemic have made it possible to support the originally face-to-face modalities in education at all levels. The teaching and learning processes require educational strategies that make it possible to take classrooms to the virtual world and be as effective as in the face-to-face setting. The objective of this work is to determine a virtual classroom design model that generates a positive effect on the motivation and performance of the participants. A descriptive methodology with quantitative approach is developed using the survey technique in two courses of first academic cycle of engineering careers at the Salesian Polytechnic University in the city of Guayaquil, during two academic periods, one from May to September 2020, another from October 2020 to February 2021, and five courses during May to September 2021. The analysis determined the relationship of the virtual classroom design, with the order, graphic line, formats, and amount of resources, with the motivation and performance of the participants. It is concluded the positive effect that the learning activities have, making them more meaningful for the student, in relation to the model of their presentation, generating an intention of appropriation and belonging on the part of the participant, due to its immediate comprehension and ease of realization. This study presents a virtual classroom design model, as well as a set of digital educational media and meaningful learning activities for virtual environments on Moodle, which have proven to be effective in their application.

Keywords: Instructional design · Technology enhanced learning · e-learning · COVID-19

1 Introduction

During the COVID-19 pandemic, it has become more evident that teaching and learning processes supported by virtual environments require models that allow content planning. [1, 2], activities and experiences from traditional classrooms to a virtual environment, aspiring to the qualities of face-to-face attendance [3, 4]. From the theory of planned

S. Berrezueta and K. Abad (Eds.): *Doctoral Symposium on Information and Communication Technologies - DSICT*, LNEE 846, pp. 21–32, 2022.
https://doi.org/10.1007/978-3-030-93718-8_3

behavior (TPB), it is stated that behavior is determined from an intention that involves the impulse to achieve an accomplishment [5]. If that impulse is attracted by the interest of the activity, finding a pleasure or challenge in doing it, rather than the reward for doing it, intrinsic motivation is evident [6, 7]. In this sense, minimizing negative states of participants in online learning systems for the development of activities will require involving strategies that improve behavior from the emotional and cognitive point of view, increasing or maintaining motivation [8, 9]. Several studies suggest that online virtual learning platforms or environments require the incorporation, by the teaching staff, of a design that tries to stimulate motivation for learning; visualizing in the virtual space, materials or digital educational media that generate confidence, attention and reflection in the participant [10–12]; as well as, self-determination to carry out the activities organized for their significance within the planned learning proposal [13, 14]. Adequate visualization of learning activities by the participant is a strategy to improve their effectiveness, optimize their efficiency and boost satisfaction with the activity [15–17].

Teaching and learning spaces during the COVID-19 pandemic have moved to online mode [18, 19], pushing researchers to focus all their efforts to generate strategies supported by ICTs tools [20, 21], enabling institutions to deal with the emerging online education from their material and human resources [22, 23]. As for the participant, in addition to carrying out activities based on the objective specified in the content, he/she must make a cognitive effort to understand the activity from the format in which it is presented [24, 25]. In other words, the cognitive and affective factors of the learner must be considered; this includes observation and the reason for carrying out the activity [26], as well as that which generates an impulse (motivation) to initiate it [27–29].

This work aims to determine a model of virtual classroom design within a Moodle environment, which generates a positive effect on the motivation and performance of participants.

2 Materials and Methods

An empirical-analytical methodology of quasi-experimental cut with quantitative approach is developed, using the survey technique with Likert-type scale, applied to three groups of students, all of them of the first level of the engineering careers of the Universidad Politécnica Salesiana in the city of Guayaquil-Ecuador. The first group is represented by 36 participants out of a population of 40 first level students who belong to the academic period from May to September 2020 (period 56); the second group are 38 participants out of 40 first level students who belong to the period from October 2020 to February 2021 (period 57); and the third group are 261 participants out of 265 first level students who belong to the period from May to September 2021 (period 58). In other words, almost the entire population involved in this study participated (through the virtual classrooms).

Three processes are developed during the research: first, meaningful learning activities are elaborated; second, the integration of the activities in the virtual environment is developed with an *ad hoc* model for the course (according to the course planning or program); third, the students' learning activities are evaluated for continuous improvement of the activity.

Elaboration of significant learning activities (First process). In the implementation of the courses, a group of professors who teach the subject in different parallels are in charge of developing the learning activities (common practice in universities [30, 31]). According to a planning that involves activities in three areas, *Learning in contact with the teacher*, *Learning by doing* and *Self-directed learning*, these activities can use resources from external programs to generate meaningful learning objects. Based on the expertise of the members of the teaching team, activities are generated outside or within the Moodle platform and then integrated into the corresponding sections (see Fig. 1).

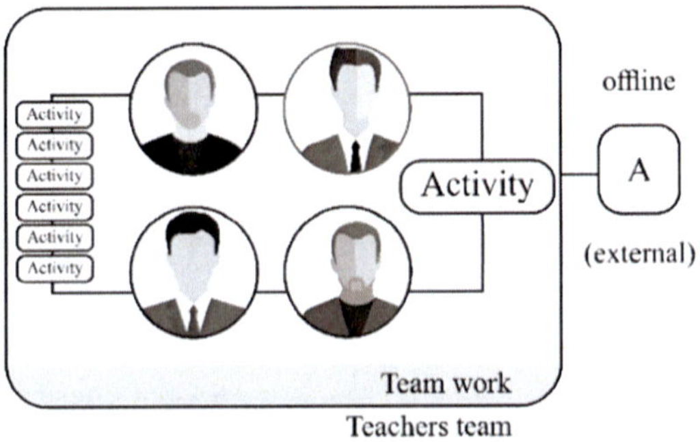

Fig. 1. Team of teachers who model the set of ad hoc content and activities and their objectives through an instructional design for the virtual learning environment.

Integration of activities in the virtual environment with an ad hoc model (Second process). Teachers incorporate digital educational media in learning activities according to the tools and technological support provided by the institution's virtual environment (Moodle), (see Fig. 2).

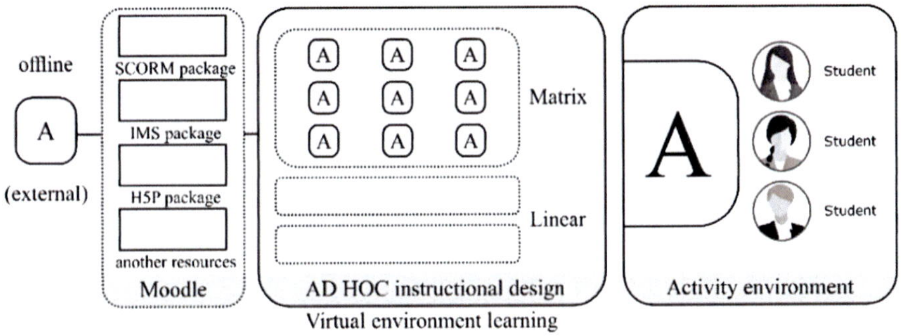

Fig. 2. Proposal for an instructional model in the virtual learning environment.

The learning platform of the Universidad Politécnica Salesiana (UPS) uses a standard template in its virtual environment that allows incorporating native hypertext language code and embedding packages of technologies allowed (SCORM, IMS, H5P, among others) by the platform version (currently Moodle 3.9). Previous work has enabled the implementation of a set of significant activities [10, 15], according to the learning process determined by the professors who teach the course, adapted from [25], (see Table 1).

Table 1. Significant learning activities (ASA) for the teaching-learning process.

Area	Sub-area	Code	Significant learning activities (ASA)
Learning in contact with the teacher	Collaborative learning	ASA-01	PP – Project of problematization proposed by the teacher in class.
		ASA-02	In-depth study of topics.
		ASA-03	Knowledge integrating work
	Assisted learning	ASA-04	Reading control - inverted classroom
		ASA-05	Tutoring - consultancy
		ASA-06	Academic WhatsApp Group
Learning by doing		ASA-07	PG – Guided Practice Application and Experimentation
		ASA-08	Peer mentoring
		ASA-09	Crowdlearning workgroups
		ASA-10	Laboratory practice
Self-directed learning		ASA-11	PI – Individual production of Works
		ASA-12	Documentary videos
		ASA-13	Virtual exhibitions, YouTuber initiative
		ASA-14	Remote communications space
		ASA-15	Preparation and training simulators

Evaluation of learning activities by students as continuous improvement (Third Process). At the beginning of each academic period, students' courses are activated so that they can access the contents. Students can access the course without the need of the teacher's accompaniment or wait for the beginning of the class period to explore the course [32]. Upon entering the virtual learning platform, during the corresponding class time, students explore the sections, within a process of setting guided by the course teacher. The availability of resources depends on the teacher's skills, as the teaching and learning process develops.

Participating students encounter in their courses the proposed matrix instructional model for significant learning activities (ASA); a model that identifies the three dimensions of interaction, passive activities, interactive activities, and experiential activities (see Fig. 3).

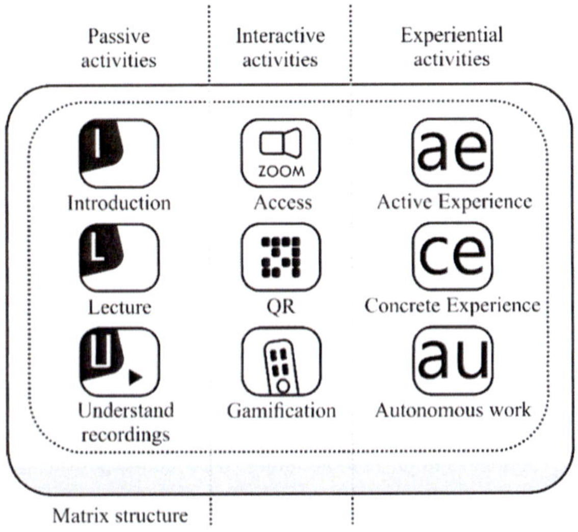

Fig. 3. Instructional matrix model of meaningful learning activities (ASA)

Matrix Model for Presentation and Access to Activities. The use of the matrix model of resource presentation has been studied in previous works [33–36] and is visible in every aspect of the elements that the learner finds in the units of study. Each dimension contains three fundamental elements of quick access for the student. In the dimension of passive activities are the elements of deepening, these are "Introduction to the unit", "Readings of documents or video", "readings supported with technologies" and the "recordings of each day" of synchronous class. In the "interactive activities" dimension, there are links to the synchronous meeting using Zoom, the link to the resources hosted in a single Google Drive repository by accessing a QR code and the third link to access the evaluation platform through games [37]. In a third dimension of experiential activities, there are the activities to be developed by the student individually during class (active experience), in groups with their assigned classmates (concrete experience) and the activity of autonomous work outside of class (Fig. 4).

Learning Outcome Products. Work submitted by students as a result of individual work in class, outside of class or in a group setting by applying crowdlearning [38] are delivered in traditional formats or in multimedia elements developed by the participants with creativity. The delivery of formats with designs of the graphic line of the subject allows students to use them and follow a similar scheme. These works maintain academic rigor in their components, objectives, activities, and conclusions to be reached.

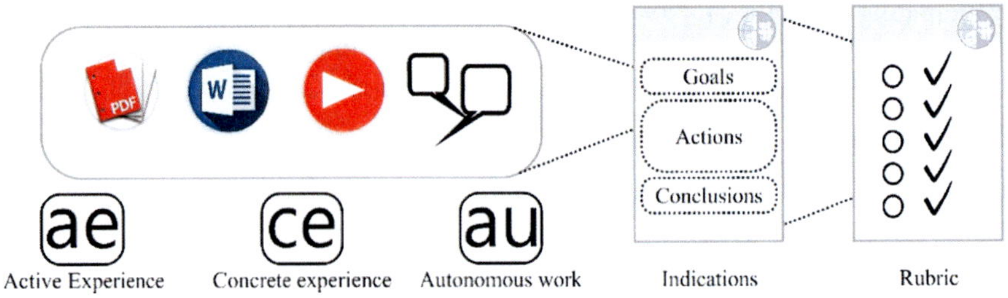

Fig. 4. Outputs of experiential learning activities according to outcome indicators and rubrics

3 Results and Discussion

The number of participants for this study is detailed in Table 2, highlighting that the number of young participants under 18 years of age is evident in the first levels of the current period, 58, the percentage of older students increases in previous periods due to the advancement of the years. A survey was conducted for each group of students using online tools and the responses are presented below.

Table 2. Sample demographics by academic period.

Academic period		58		57		56	
Measure	Items	Frecuency	Percent	Frecuency	Percent	Frecuency	Percent
Gender	Female	77	30%	10	26%	8	22%
	Male	184	70%	28	74%	28	78%
Age	<=18	90	35%	2	5%	1	3%
	19–21	131	50%	20	53%	16	44%
	22–25	31	12%	12	32%	16	44%
	>=26	9	3%	4	10%	3	9%

In each academic period, the percentage of male students is higher than that of female students. In fact, it is evident that in the groups of engineering students the majority are men, but the growth of interest in engineering careers by women is notable.

The participating groups that had the experience of using the experimental design model in their virtual environment courses were asked, "To what extent does the organization of the different elements of the virtual classroom (sections, contents and links) contribute to their motivation for learning the subject?", (see Fig. 5).

The students responded to a survey with the options *a lot of*, *enough*, *something*, *little*, *nothing*; obtaining the response *a lot of*, in period 56 of 64%, in period 57 of 76% and for period 58 (still valid for this research work) it is at 41%. It is understood that the application of the model in period 57 had improvements after the experience in period 56 (six months later). Currently, in the middle of period 58, a high percentage of contribution from the application of the model is noted. If we add to this the percentage of *enough*, these results exceed 83% on average.

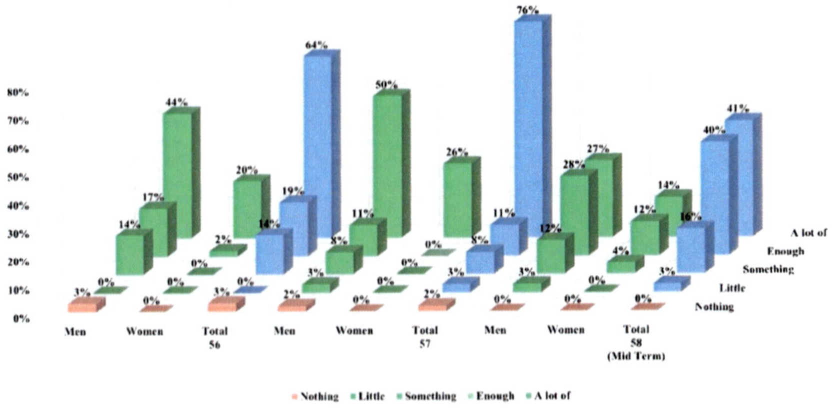

Fig. 5. To what extent the organization of the different elements of the virtual classroom (sections, contents, and links) contribute to their motivation for learning the subject.

Students were also asked "To what extent do the graphic line, formats, images, labels and visual elements that you find in the virtual classroom, contribute to your motivation for learning the subject" (see Fig. 6).

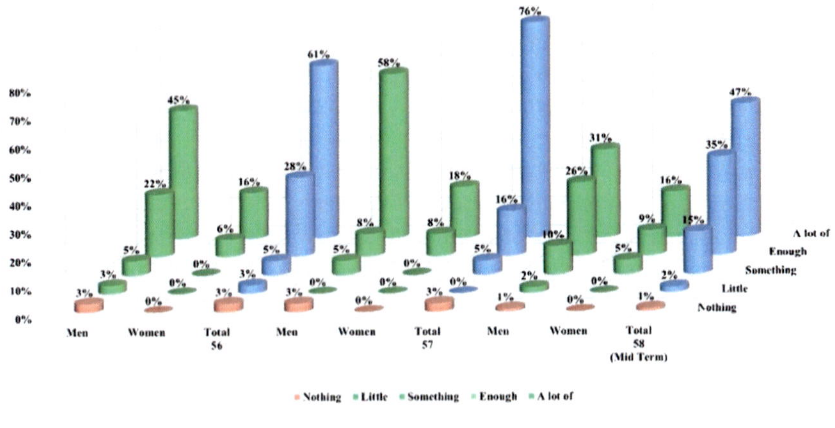

Fig. 6. To what extent the graphic line, formats, images, labels, and visual elements that you find in the virtual classroom, contribute to their motivation for learning the subject.

In this question a slightly higher result was obtained in general, the response *a lot of*, in period 56 was 61%, in period 57 76% and for period 58 (still valid for this research work) it is at 47%. The improved resources for period 57 had a higher contribution to learning motivation. If we add to this the percentage of *enough*, these results exceed 87% on average.

Students were also asked "To what extent have the number of resources you have found in the virtual classroom been useful for learning the respective units planned in the course", (see Fig. 7).

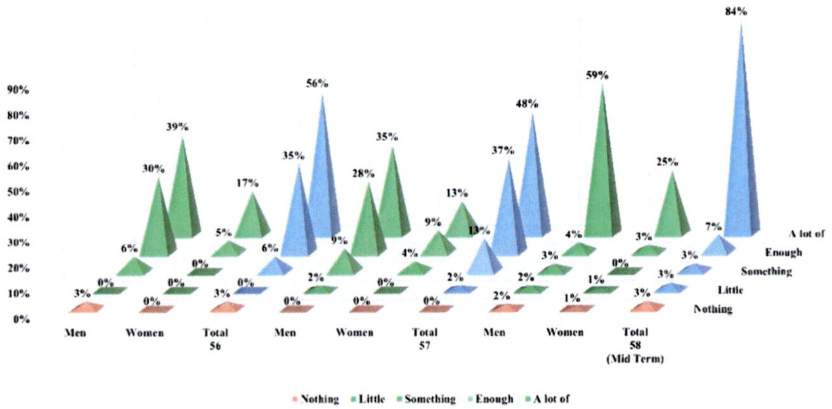

Fig. 7. To what extent, the number of resources that you have found in the virtual classroom, have been useful for learning the respective units planned in the subject.

Obtaining a superior result in general, the response *a lot of*, in period 56 was 56%, in period 57 it was 48% and for period 58 (still valid for this research work) it is at 84%. The improved resources for the current period are more useful for learning the respective units planned in the course. If to this we integrate the percentage of *enough*, these results reach 89% on average. It is understood that the experience in the two periods has achieved a value of 84% of usefulness in the middle of period 58. It is remarkable to note that each semester with the feedback of the students themselves has been possible to have a continuous improvement of the digital educational resources.

Finally, we present the set of ten (10) educational digital media (MED) and the details of their descriptions, applied in the virtual environments during the three academic periods (see Table 3), periods that took place during the COVID-19 pandemic [39].

The analysis of the results determines that the relationship of the design of the virtual classroom, with the order, the graphic line, document formats according to the graphic scheme and its structured contents, and the number of appropriate resources, are related to the motivation and performance that students have for their participation, development, and achievement with a sense of belonging to its attainment.

Table 3. Digital educational media for virtual learning environments.

Code	Description of the resource	Code	Description of the resource
MED-01	Welcome video to the course by the teacher, YouTube is used as a playback channel, content prepared in PowerPoint with insertion of images and musical media, available at https://youtu.be/VmArGOLynhE	MED-06	Video lecture elaborated by the teacher for additional indications to the topic, with the use of iSpring for PowerPoint
MED-02	Course materials in digital format with defined graphic line	MED-07	Documentary videos that address related topics, using streaming platforms
MED-03	Access to the Virtual Library for consultation of digital documents, as well as indexed journals	MED-08	Use of additional third-party software and complementary communication tools
MED-04	Use of remote tools for group tasks	MED-09	Simulator of evaluations by questionnaire of questions
MED-05	External access to the repository of subject resources, http://bit.ly/AL53-UPS with the use of the Google Drive repository	MED-10	Visibility of grades in honor roll format

4 Conclusions

The effectiveness of the performance of the activities explained by the perception of affinity with the task; confidence in its completion; and satisfaction with the learning outcome to be achieved, contributed positively to the participant's motivation. The participants, being motivated, are more likely to acquire a high commitment to start and finish a task, as well as to reach a better understanding of it.

It is possible to conclude that an adequate design of the learning activities, configuring a virtual classroom, makes them more stimulating for the student and this influences the effective completion of the proposed tasks and the associated performance results. This depends on the presentation (order, graphic line, formats, and quantity of resources) that generate an intention of appropriation and belonging on the part of the participant; promoting an immediate understanding of the meaning of the activities and resources; and the desire to complete their realization.

Acknowledgments. We thank the young people of the times involved in this work, the Universidad Politécnica Salesiana for the support received and the Vice-Rectorate for Research of the university for the constant monitoring and accompaniment.

References

1. Abd Elrahman, M.G.S., Hassan, A., Gunied, H., Al Saedi, K.M.A., Mohammed, F.A.: The usage of E-learning among mass communication students. In: Musleh Al-Sartawi, A.M.A. (ed.) The Big Data-Driven Digital Economy: Artificial and Computational Intelligence. SCI, vol. 974, pp. 47–57. Springer, Cham (2021). https://doi.org/10.1007/978-3-030-73057-4_4
2. Ashrafi, A., Zareravasan, A., Rabiee Savoji, S., Amani, M.: Exploring factors influencing students' continuance intention to use the learning management system (LMS): a multi-perspective framework. Interact. Learn. Environ. (2020). https://doi.org/10.1080/10494820.2020.1734028
3. Wang, R., Chen, L., Solheim, I.: Modeling dyslexic students' motivation for enhanced learning in E-learning systems. ACM Trans. Interact. Intell. Syst. **10** (2020). https://doi.org/10.1145/3341197
4. Lin, C.-L., Jin, Y.Q., Zhao, Q., Yu, S.-W., Su, Y.-S.: Factors influence students' switching behavior to online learning under COVID-19 pandemic: a push–pull–mooring model perspective. Asia-Pac. Educ. Res. **30**(3), 229–245 (2021). https://doi.org/10.1007/s40299-021-00570-0
5. Kim, E.J., Kim, J.J., Han, S.H.: Understanding student acceptance of online learning systems in higher education: application of social psychology theories with consideration of user innovativeness. Sustain **13**, 1–14 (2021). https://doi.org/10.3390/su13020896
6. Miller, K.A., Deci, E.L., Ryan, R.M.: Intrinsic motivation and self-determination in human behavior. Contemp. Sociol. **17**, 253 (1988). https://doi.org/10.2307/2070638
7. Torres-Toukoumidis, A., Carrera, P., Balcazar, I., Balcazar, G.: Descriptive study of motivation in gamification experiences from higher education: systematic review of scientific literature. Univers. J. Educ. Res. **9**, 727–733 (2021). https://doi.org/10.13189/ujer.2021.090403
8. Yamani, H.A.: A Conceptual framework for integrating gamification in eLearning systems based on instructional design model. Int. J. Emerg. Technol. Learn. **16**, 14–33 (2021). https://doi.org/10.3991/ijet.v16i04.15693
9. Ma, L., Lee, C.S.: Evaluating the effectiveness of blended learning using the ARCS model. J. Comput. Assist. Learn. jcal.12579 (2021). https://doi.org/10.1111/jcal.12579
10. Llerena-Izquierdo, J., Ayala-Carabajo, R.: University teacher training during the COVID-19 emergency: the role of online teaching-learning tools. In: Rocha, Á., Ferrás, C., López-López, P.C., Guarda, T. (eds.) ICITS 2021. AISC, vol. 1331, pp. 90–99. Springer, Cham (2021). https://doi.org/10.1007/978-3-030-68418-1_10
11. Saleem, A.N., Noori, N.M., Ozdamli, F.: Gamification applications in E-learning: a literature review. Technol. Knowl. Learn., 1–21 (2021). https://doi.org/10.1007/s10758-020-09487-x
12. Al Ghawail, E.A., Ben Yahia, S., Alrzini, J.R.: Gamification model for developing E-learning in Libyan higher education. In: Uskov, V.L., Howlett, R.J., Jain, L.C. (eds.) Smart Education and e-Learning 2021. SIST, vol. 240, pp. 97–110. Springer, Singapore (2021). https://doi.org/10.1007/978-981-16-2834-4_9
13. Bernik, A.: Gamification framework for E-learning systems in higher education. Teh. Glas. **15**, 184–190 (2021). https://doi.org/10.31803/tg-20201008090615
14. Echeverría Rodríguez, L., Cobos, R.: The impact of a motivation booster in blended learning experiences in computer science education. Rev. EIA. **18**, 1–12 (2021). https://doi.org/10.24050/reia.v18i35.1320
15. Llerena-Izquierdo, J., Atiaja-Balseca, L.: Gamification within the learning evaluation process using Ardora at the Salesian Polytechnic University (Guayaquil, Ecuador). In: Botto-Tobar, M., Montes León, S., Camacho, O., Chávez, D., Torres-Carrión, P., Zambrano Vizuete, M. (eds.) ICAT 2020. CCIS, vol. 1388, pp. 139–150. Springer, Cham (2021). https://doi.org/10.1007/978-3-030-71503-8_11

16. Wang, Y., Jiang, Y.L.: Research on learning behavior based on ARCS learning motivation model. Adv. Sci. Technol. **105**, 356–363 (2021). https://doi.org/10.4028/www.scientific.net/ast.105.356
17. Goksu, I., Islam Bolat, Y.: Does the ARCS motivational model affect students' achievement and motivation? Meta-Anal. Rev. Educ. **9**, 27–52 (2021). https://doi.org/10.1002/rev3.3231
18. Bacher-Hicks, A., Goodman, J., Mulhern, C.: Inequality in household adaptation to schooling shocks: covid-induced online learning engagement in real time. J. Public Econ. **193**, 104345 (2021). https://doi.org/10.1016/j.jpubeco.2020.104345
19. Aditya, D.S.: Embarking digital learning due to COVID-19: are teachers ready? J. Technol. Sci. Educ. **11**, 104–116 (2021). https://doi.org/10.3926/jotse.1109
20. De La Serna-Tuya, A.S., Salgado-Gutierrez, D., Ochoa-Garcia, J., Mora-Lopez, A.F., Garcia-Bejar, L.: Digital Tools for virtual courses for university teachers affected by COVID-19. In: Proceedings - 10th International Conference on Virtual Campus, JICV 2020. Institute of Electrical and Electronics Engineers Inc. (2020). https://doi.org/10.1109/JICV51605.2020.9375698
21. Urakova, S.: ICT and the new teaching and learning spaces: from new pedagogical models to virtual teaching and learning environments. Журнал иностранных языков и лингвистики 2 (2021)
22. Kahambing, J.G.S.: Time to reopen schools: COVID-19, health disparity and education. J. Public Health (Bangkok) (2021). https://doi.org/10.1093/pubmed/fdab205
23. Azubuike, O.B., Adegboye, O., Quadri, H.: Who gets to learn in a pandemic? Exploring the digital divide in remote learning during the COVID-19 pandemic in Nigeria. Int. J. Educ. Res. Open. **2**, 100022 (2021). https://doi.org/10.1016/j.ijedro.2020.100022
24. Hudlicka, E.: To feel or not to feel: the role of affect in human-computer interaction. Int. J. Hum. Comput. Stud. **59**, 1–32 (2003). https://doi.org/10.1016/S1071-5819(03)00047-8
25. Llerena, J., Ayala-Carabajo, R.: Significant learning activities (ASA) in the modality of face-to-face studies with integration of virtual educational media in Engineering careers. In: Proceedings of the 15th Latin American Conference on Learning Technologies, LACLO 2020, pp. 1–9 (2020). https://doi.org/10.1109/LACLO50806.2020.9381134
26. Keller, J.: Motivational design of instruction. Instr. -Design Theor. Model. **1**, 386–434 (1983)
27. Okey, J.R., Santiago, R.S.: Integrating instructional and motivational design. Perform. Improv. Q. **4**, 11–21 (1991). https://doi.org/10.1111/j.1937-8327.1991.tb00500.x
28. Lock, J., Lakhal, S., Cleveland-Innes, M., Arancibia, P., Dell, D., De Silva, N.: Creating technology-enabled lifelong learning: a heutagogical approach. Br. J. Educ. Technol. (2021). https://doi.org/10.1111/bjet.13122
29. Alamri, H.A., Watson, S., Watson, W.: Learning technology models that support personalization within blended learning environments in higher education. TechTrends **65**(1), 62–78 (2020). https://doi.org/10.1007/s11528-020-00530-3
30. Yang, W., Li, Q., Wang, X.: The effect of powerpoint theme color on learners' learning effectiveness and emotions. In: 2021 IEEE 24th International Conference on Computer Supported Cooperative Work in Design (CSCWD), pp. 510–512. IEEE (2021). https://doi.org/10.1109/CSCWD49262.2021.9437633
31. Nurminen, M., Leppänen, L., Väätäjä, H., Ihantola, P.: Search of the emotional design effect in programming revised. In: Lavoué, É., Drachsler, H., Verbert, K., Broisin, J., Pérez-Sanagustín, M. (eds.) EC-TEL 2017. LNCS, vol. 10474, pp. 434–440. Springer, Cham (2017). https://doi.org/10.1007/978-3-319-66610-5_39
32. Llerena Izquierdo, J.: Uso de AVAC en la Universidad Politécnica Salesiana. https://dspace.ups.edu.ec/handle/123456789/10913
33. Kiewra, K.A., Kim, S. Il, Risch, N., Mayer, R.E., Dubois, N.F., Christensen, M.: Effects of advance organizers and repeated presentations on students' learning. J. Exp. Educ. **65**, 147–159 (1996). https://doi.org/10.1080/00220973.1997.9943789

34. Imhof, B., Scheiter, K., Edelmann, J., Gerjets, P.: How temporal and spatial aspects of presenting visualizations affect learning about locomotion patterns. Learn. Instr. **22**, 193–205 (2012). https://doi.org/10.1016/j.learninstruc.2011.10.006
35. Hancock-Niemic, M.A., Lin, L., Atkinson, R.K., Renkl, A., Wittwer, J.: Example-based learning: exploring the use of matrices and problem variability. Education Tech. Research Dev. **64**(1), 115–136 (2015). https://doi.org/10.1007/s11423-015-9403-8
36. Colliot, T., Jamet, É.: Improving students' learning by providing a graphic organizer after a multimedia document. Br. J. Educ. Technol. **52**, 252–265 (2021). https://doi.org/10.1111/bjet.12980
37. Bouman, M.: A strange kind of marriage: the challenging journey of entertainment-education collaboration. In: Frank, L.B., Falzone, P. (eds.) Entertainment-Education Behind the Scenes, pp. 61–83. Springer, Cham (2021). https://doi.org/10.1007/978-3-030-63614-2_5
38. Llerena-Izquierdo, J., Barcia-Ayala, O., Ayala-Carabajo, R.: Faculty training through crowdlearning for emerging online education. In: 2020 IEEE ANDESCON, pp. 1–7 (2020). https://doi.org/10.1109/ANDESCON50619.2020.9272103
39. Lorenzo Moledo, M., Quiroga-Carrillo, A., Santos Rego, M.A.: La educación en Red: realidades diversas, horizontes comunes, XVII Congreso Nacional y IX Iberoamericano de Pedagogía, Santiago de Compostela, 7–9 de julio de 2021, libro de resúmenes. Universidad de Santiago de Compostela, Servicio de Publicaciones e Intercambio Científico (2021). https://doi.org/10.15304/cc.2021.1393

Modeling of the Guangarcucho Municipal Wastewater Treatment Plant Using WEST, Cuenca-Ecuador

Paola Duque-Sarango[1,2](✉) and Verónica Pinos[2]

[1] Universidad Politécnica Salesiana, Cuenca, Ecuador
pduque@ups.edu.ec
[2] Departamento de Recursos Hídricos y Ciencias Ambientales, Universidad de Cuenca, Cuenca 010203, Ecuador
veronica.pinos@ucuenca.edu.ec

Abstract. The present work applied the ASM1 simulation model for the wastewater treatment system of Guangarcucho plant (G-WWTP) in the city of Cuenca. The main objective was to generate conclusions about parameters that determine the quality of the effluent, and to have a first approximation of the operating processes in the plant. In order to achieve this, a work routine was developed with the WEST software, data on pollutant concentration and flow rates from a year of daily records (year 2018–2019) was taken at the entrance of the Ucubamba current wastewater treatment plant (U-WWTP), which were provided by ETAPA-EP, while kinetic parameters were taken from those used for designing G-WWTP plant. Steady-state simulations were performed to later analyze dynamic-state simulations. The topological configuration was structured by biological reactors, a dissolved oxygen control model, a secondary settlers model, combiners and flow dividers, among the main elements. In addition, a basic sensitivity analysis was performed on variables that can be manipulated at the macro level in the plant: dissolved oxygen concentration (DO) and internal sludge recirculation (Qlodos). It was found that internal sludge recirculation influences positively on the effluent quality, causing concentration of organic matter (BOD) and decrease of the suspended solids (TSS) to decrease, while the concentration of dissolved oxygen higher or lower than 2.5 mg/L, does not have a greater influence on the quality variation of the effluent.

Keywords: Wastewater treatment plant · Simulation model · WEST software

1 Introduction

Wastewater treatment plants (WWTP) play a fundamental role in the water cycle, producing a liquid effluent of adequate quality to be returned to the natural surface, achieving a minimum impact on the environment and humans health [1–3].

The conventional most widely used wastewater treatment technologies include chemical, physical and biological treatments; many of these are effective and depend on the

S. Berrezueta and K. Abad (Eds.): *Doctoral Symposium on Information and Communication Technologies - DSICT*, LNEE 846, pp. 33–46, 2022.
https://doi.org/10.1007/978-3-030-93718-8_4

type of pollutant to be treated. The biological treatment consists of biodegradable pollutants elimination from the water by microorganisms, as a consequence of organic matter and nutrients consumption from the residual water that generates biomass (mud or sludge) that requires subsequent extraction [1, 4].

Currently the design and operation of WWTPs are validated using simulators, which incorporate complex mathematical models that include algebraic and differential equations to represent the main physicochemical and biological phenomena in the water treatment process [4, 5]. Modeling is an inherent part of the design and operation of a wastewater treatment system. A suitable mathematical model for biological wastewater treatment must consider the essential phenomena (biological and physicochemical) and the main variables (biomass, substrate, dissolved oxygen, etc.) of the process to guarantee a reliable representation, for a correct decision-making [6–8].

In this context, IWA (International Water Association) has developed a series of activated sludge models, considered as good solutions to correlate the complexity of activated sludge processes and the prediction of biological treatment efficiency under dynamic conditions. The ASM1 model was the first model developed, it includes carbon oxidation under aerobic and anoxic conditions, as well as nitrification and denitrification [7, 9].

These mathematical models can be used during the design, operation, and optimization of a wastewater treatment system. To do so, a good software tool is indispensable. WEST is a general modelling and simulation environment and it can, together with a model base, be used for this task. The model base presented here is specific for biological wastewater treatment and is written in MSL-USER [1].

This research develops the system modeling of the future G-WWTP treatment plant, applying the Activated Sludge Model ASM1, using the simulation software WEST® [9], with the aim of generating conclusions about the parameters that determine the quality of the effluent, and have a first approximation of the operating processes in the plant.

1.1 Description of the G-WWTP

In the city of Cuenca-Ecuador, there is Ucubamba Wastewater Treatment Plant (U-WWTP) that has been in operation since 1999, and processes an average flow of 1.8 $m^3 \cdot s^{-1}$. This plant is divided into two identical flow lines, each of which comprises an aerobic pond, a facultative pond and a maturation pond to treat both municipal and industrial wastewater [10]. The construction of a new wastewater treatment plant (G-WWTP) is currently planned. This new plant will increase current wastewater treatment capacity and include flow from suburban areas that will be connected to the urban sewer system [11, 12]. The G-WWTP was designed to remove organic carbon with activated sludge technology [11].

1.2 Description of the Activated Sludge Model No.1 (ASM1)

The matrix format proposed in the ASM1 model compactly expresses the biochemical reactions that take place in the described system using a stoichiometry matrix and a kinetic vector, which facilitate the understanding of the model and the monitoring of the interactions between components. This model incorporates all transformations and components necessary to describe hydrolysis, organic matter biodegradation under aerobic and anoxic conditions (denitrification) and nitrification process in which ammonia nitrogen is oxidized to nitrates [13, 14]. Figure 1 shows a diagram of the components and transformations that are considered in the ASM1 model.

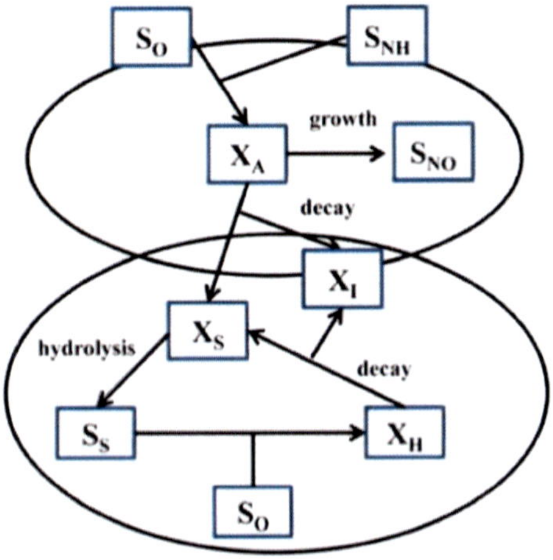

Fig. 1. Component and processes included in the ASM1 model [9]

Soluble and particulate organic matter is represented by the components S_S and X_S respectively. From S_S component, the growth of heterotrophic bacteria (X_H) occurs under aerobic and anoxic conditions. X_A component represents the populations of autotrophic bacteria that, in a simplified way, completely oxidize ammonia nitrogen (S_{NH}) to nitrates (S_{NO}) in a single step. The disappearance of bacteria X_H as well as X_A is described by lysis-regeneration concept in which the components obtained in the disappearance of the bacteria are X_S and inert matter represented by X_I component [9, 13]. The unit of measure for these components is the Chemical Oxygen Demand (COD) (see Fig. 1).

2 Simulation Routine

In this study, the ASM1 modeling of G-WWTP plant was carried out following the procedure described in Fig. 2, simulations were carried out in steady state and then in dynamic state, using as starting data, concentration database of pollutants and flow rates of one year of daily records (year 2018–2019) of (U-WWTP) and kinetic parameters used in the new plant design.

2.1 Topology in WEST 2012®

The topology was built based on the flow diagram of the G-WWTP plant, which consists of an anoxic-aerobic biological treatment unit, each one includes aeration tanks of 5000 m^3 and 11020 m^3 (process "A/O2"), a secondary settling tank (3416 m^3), the bottom flow of this tank flows back to the anoxic tank at a rate of 1.5 (3000 mg/L)

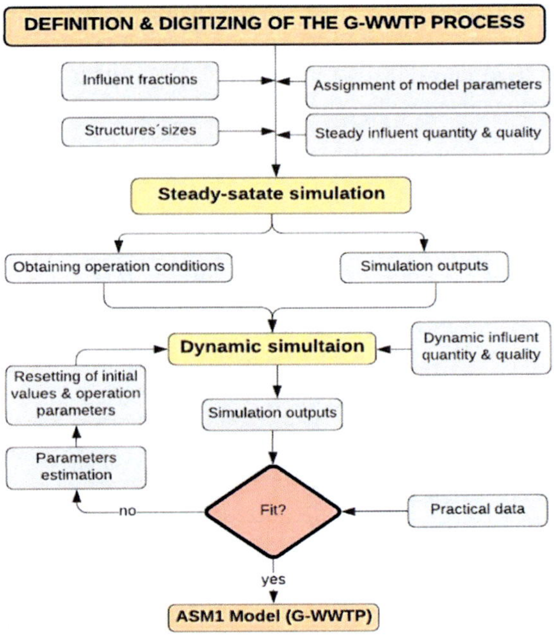

Fig. 2. ASM1 modeling and simulation procedure for G-WWTP using software WEST 2012 ®

to supply nutrient requirements of bacteria and maintain a sufficient concentration of solids. For the process operation, dissolved oxygen concentration (DO) was established equal to 2.5 mg/L (Fig. 3).

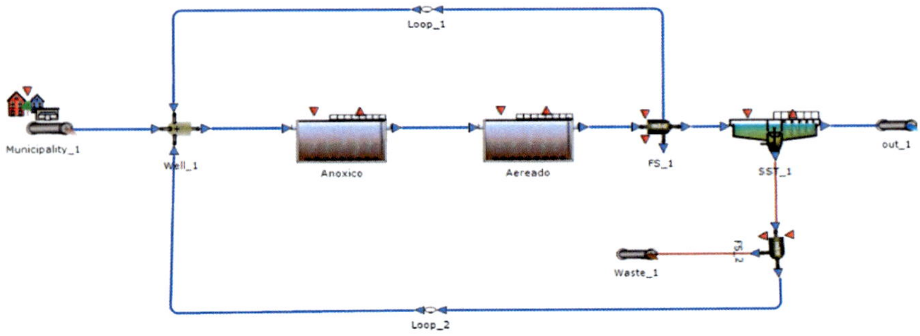

Fig. 3. G-WWTP flow chart in WEST 2012®

The general information of Sub-models used in the WEST simulator is shown in Table 1.

Table 1. General information of the units used in the WEST 2012 ®

Equipment/process	Scheme	WEST model
WWTP input	Municipality_1	Influent_Municipality_1
Two-stream divider	Well_1	TwoCombiner
Anoxic reactor	Anóxico	FixVolumeASU
Aerobic reactor	Aereado	FixVolumeASU
Secondary settler	SST_1	Takacs_SVI
WWTP effluent	out_1	Effluent_out_1
Sludge outlet	Waste_1	Effluent_Waste_1

2.2 Initial Parameters for the Model

Table 2 shows the average data of U-WWTP influent wastewater characteristics that was used in static simulation. The default values that WEST has incorporated on kinetic coefficients were used, which are: degradation kinetic terms and production of species derived from biological degradation, integral time constants and proportional gains in aeration controls, among others. These parameters have been obtained from a bibliographic compilation of various treatment plants according to the authors [9].

Table 2. Characteristics of influent wastewater in the U-WWTP (2018–2019)

	Q-Water	COD	BOD	TSS	TKN	TP
	m^3/s	mg/L	mg/L	mg/L	g/m^3	g/m^3
Values	1,771 ± 0,064	316,0 ± 179,7	222,6 ± 101,2	231,2 ± 141,0	27,2 ± 6,7	3,8 ± 1,0

Aerobic Reactors Model. The “FixVolumeASU” model was used due to the availability of information that allowed the simulation. This model describes an ideally mixed activated sludge tank with the constant volume of Qout = Qin (Fig. 4).

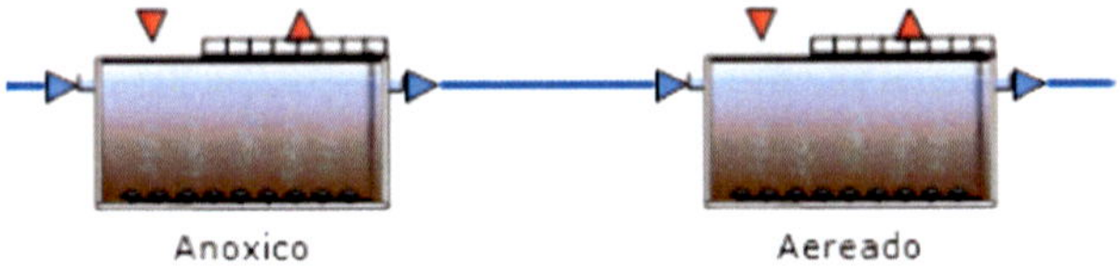

Fig. 4. Biological reactors at WEST

Secondary Settling Model. The Takacs_SVI model was used for settlers, which is an extension of the original Takacs model, and is based on estimating sedimentation by measuring the sludge volumetric index (SVI) (Eq. 1). Through experimental tests, it has been discovered that sedimentation rate of the sludge bed is a non-linear function of solids concentration in the system [15]. Consequently, sedimentation flow occurs due to gravity sedimentation and material flow (both up and down) (Fig. 5).

$$r_H = \frac{0.148 + 0.0021 * SVI}{1000} \tag{1}$$

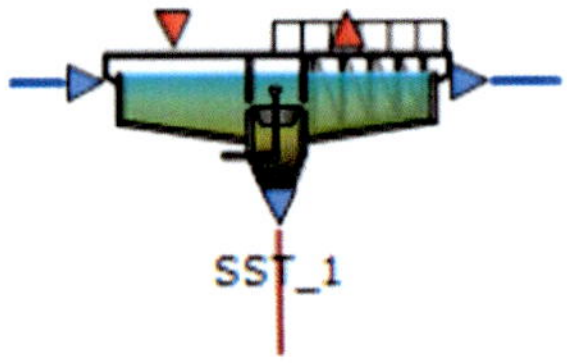

Fig. 5. Secondary settler in WEST ®

Combiner and Flow Divider Models. Flow combiners and dividers are based on mass balances without accumulation processes. In the case of combiners, the model does not require initializations since input streams determine output stream regardless of tributary inputs number.

$$Q_{out} = \sum\nolimits_{i=1}^{N} F_i \tag{2}$$

Q_{out} = Effluent flow m^3/d
N = number of inflows to the combiner
Fi = inflow flow at the i-th inlet

In case of dividers, there are two options: Flow division or Flow percentage fractionation. Initialization of each one of these equipment's is constituted by the effluent flow amount or by the effluent flow fraction to leave the equipment (Fig. 6).

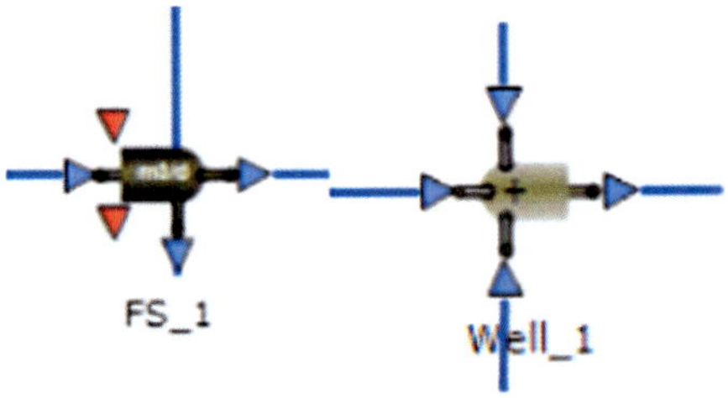

Fig. 6. Flow divider and combiner in WEST ®

3 Results and Discussion

The results of the simulation in steady state of the soluble and suspended parameters in concentration units (mg/L) are shown in Fig. 7a and 7b. It is observed that there is an adequate simulation since parameters reach stability at iteration from 25 to 30. Moreover, there is a logical explanation to the simulation; concentration of heterotrophic bacteria (X_BA) increases as soluble substrate is consumed (S_S), wich is used for biomass formation, nitrogen (ammonium) (S_NH) that they consume as a nutrient decrease as biomass increases. Additionally, new autotrophic bacteria (X_BA) also consume nitrogen, which further decreases ammonia. All this contributes to an increase of suspended solids concentration in the reactor (X_TSS).

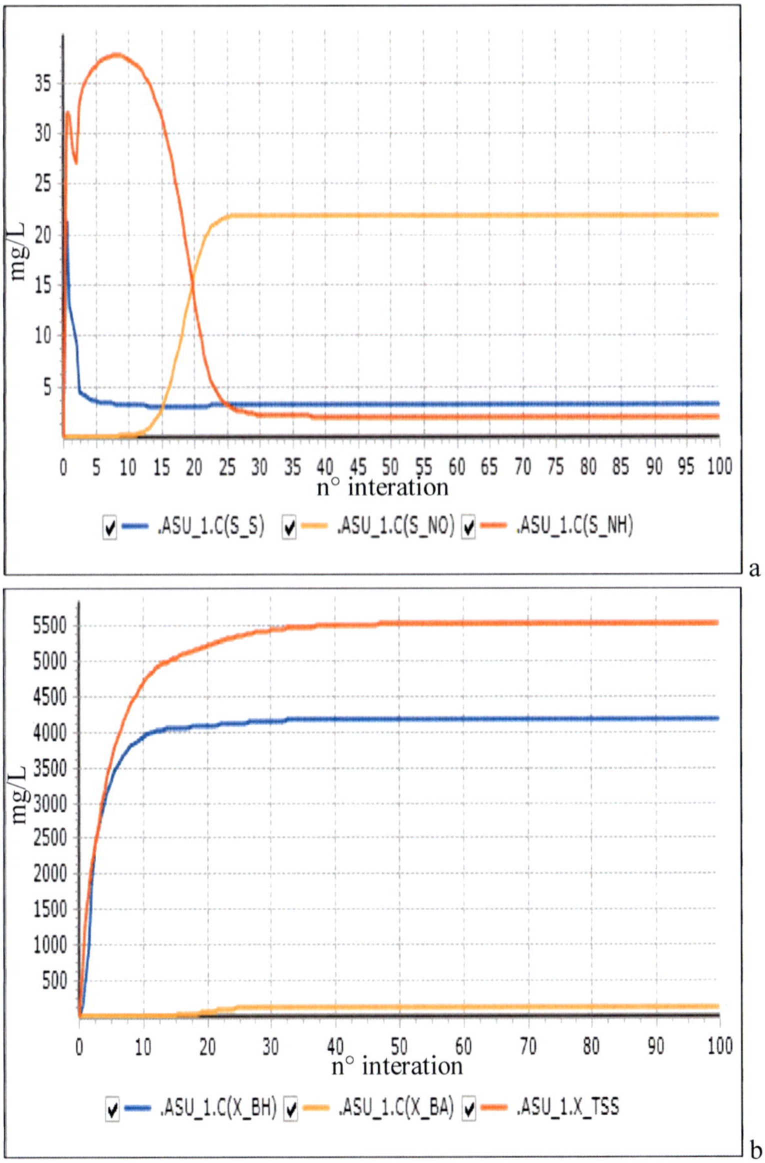

Fig. 7. a) Simulation of the soluble parameters (S_S: soluble substrate, S_NO: soluble nitrogen, S_NH: Ammonium) and b) suspended (X_BH: heterotrophic bacteria, X_BA: autotrophic bacteria, X_TSS: total suspended solids) in the biological reactor state stationary, concentration (mg/L).

The results of the simulation under dynamic conditions for a simulation time of 30 days are shown in Fig. 8.

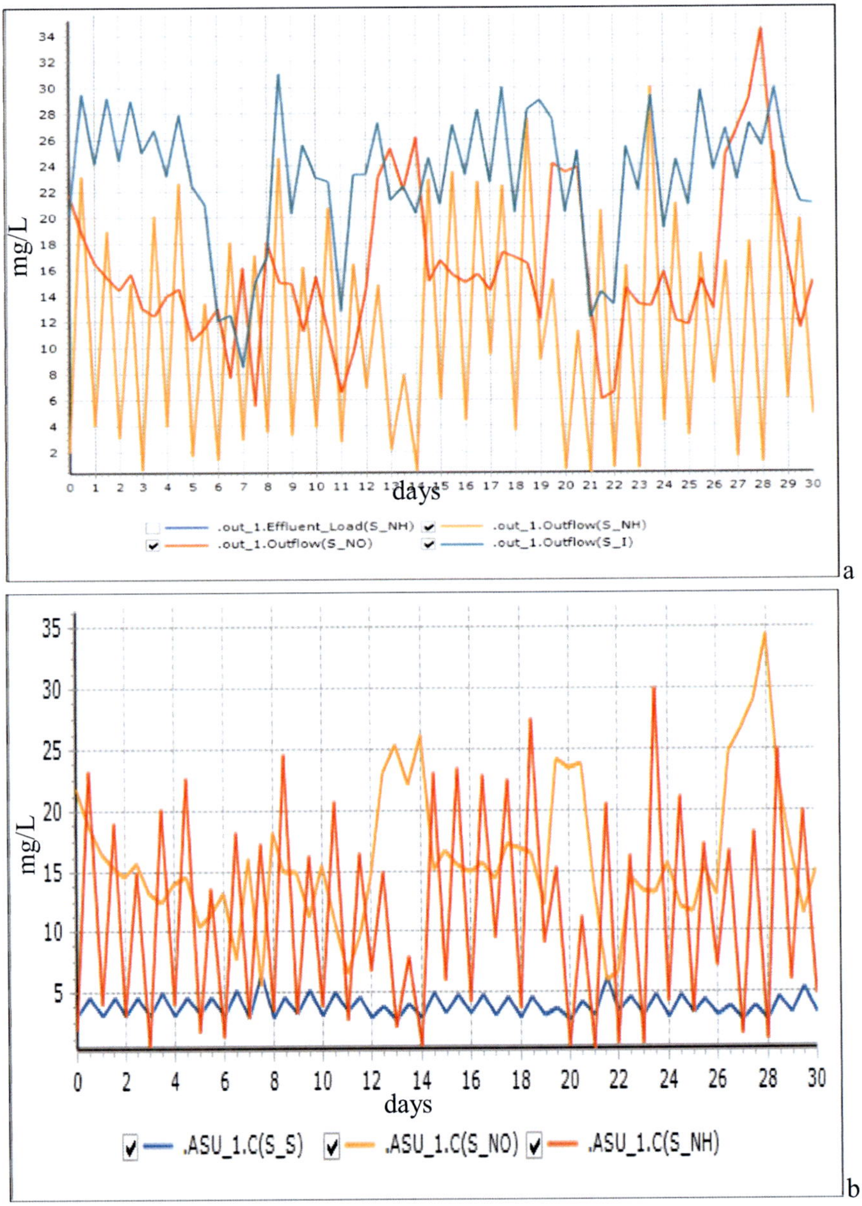

Fig. 8. Behavior of soluble (a) and suspended (b) parameters in biological reactors, under dynamic conditions, simulation 30 days, concentration (mg/L).

Figure 9 shows the behavior of organic matter parameters (COD), total nitrogen (TKN), and suspended solids concentration (TSS). From the results obtained in initial concentrations, removal percentage of these three parameters was calculated, it is evidenced that there is an important removal for COD concentration and TSS, while it is low for TKN (48%), this is due to the ASM1 model chosen for the simulation is clearly conceived to eliminate organic matter.

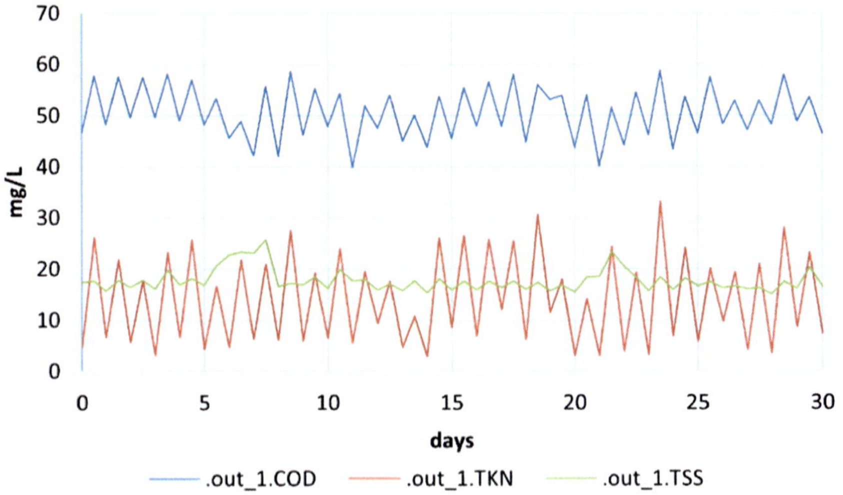

Fig. 9. Results of the G-WWTP simulation for the concentration of organic matter (COD), total nitrogen (TKN) and suspended solids (TSS).

Table 3 shows the comparison of values used in the model (influent) vs. simulation results (effluent) values and % removal. The units designed under these initial conditions are complying with minimum discharge requirements, according to Ecuadorian Legislation, which in case of BOD should not exceed 250 mg/L, for TSS can not be greater than 100 mg/L and for TKN less at 15 mg/L.

Table 3. Comparison of values used in the model (influent) vs. simulation results (effluent) values and % removal.

	COD		TSS		TKN	
	g/m^3		g/m^3		g/m^3	
Influent	316,00	±179,7	231,20	±141	27,20	±6,7
Effluent	50,55	±5,20	17,73	±2,19	14,05	±9,05
% removal	84%		92%		48%	

3.1 Sensitivity Analysis

To evaluate the sensitivity in the effluent quality, two parameters were taken. The parameters that determine the treatment process operation and the effluent quality are: 1) dissolved oxygen concentration (DO) in the aerobic reactor and 2) the internal recirculation of sludge from the settler secondary to the biological process.

The DO conditions metabolic processes of microorganisms in the biological reactor, as a design criteria, a value of 2.5 mg/L is recommended, however, one concern was to know if increasing or decreasing this value would be sensitive to generate a significant result with respect to the initial simulation routines. Regarding the internal recirculation of sludge, it can have an effect on the amount of active biomass that can return to the biological reactor, in such a way, the general hypothesis would be the greater the recirculation of sludge, the greater the amount of biomass, and the greater consumption of organic matter and nutrients, improving the characteristics of the effluent. It was evaluated what could happen if this concentration would be different from its recommended value of 3000 mg/L.

Therefore, the concentration of organic matter and suspended solids in the effluent was evaluated, varying dissolved oxygen in the aerobic reactor in a range of 1.7 to 2.5 mg/L and the sludge internal recirculation in a range of 1000 to 5000 mg/L.

The dissolved oxygen concentration is a parameter operated in the secondary treatment (aerobic reactor with biology activity) and it is necessary to analyze its sensitivity considering its influence on the microorganism's activity. From the results, it is observed that BOD is not highly changeable when oxygen concentration varies in a range of 1.5 to 2 mg/L, values obtained are close to each other. However, with a concentration of 2.5 mg/L, a drastic decrease in BOD concentration is observed, improving the quality of the effluent about the organic matter concentration (Fig. 10-a). Similarly, in the total suspended solids concentration (Fig. 10-b) there are no significant variations depending on the dissolved oxygen concentration, however, concentration of 2.5 mg/L is still important since with this value, the concentration of suspended solids in the effluent significantly decreases. The obtained can be explained because as there is more dissolved oxygen (2.5 mg/L), the activity and growth of microorganisms increases, which generates a greater removal of BOD (decrease in concentration) and an increase of TSS amount due to their death in the end of their cycle.

Sludge recycling (secondary sludge that does not recirculate) refers to the sludge amount that re-enters in the biological reactor to improve water quality. Consequently, it can be observed that as a greater quantity of sludge is re-entered in the aerobic reactor, effluent quality improves, this is because re-entry sludge promotes bacterial activity that degrades the organic matter in the water.

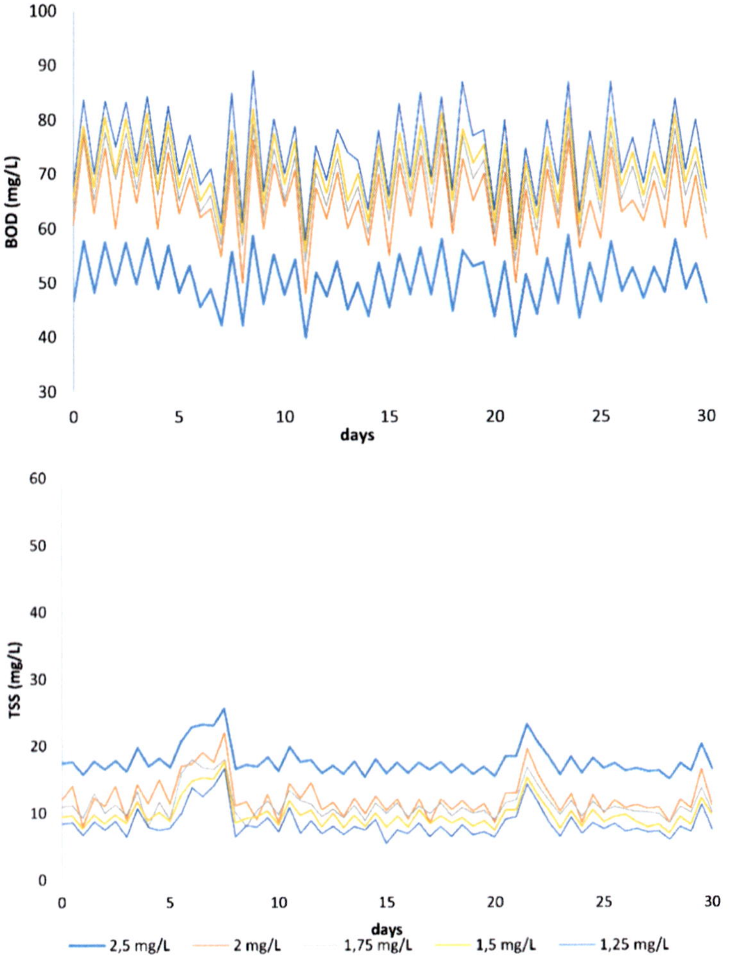

a b

Fig. 10. a) BOD (mg/L) effluent and b) TSS (mg/L) effluent, depending of the DO concentration in the biological reactor.

It is observed that sludge variation that is not recirculated shows a considerable linear proportional or positive correlation with the BOD (Fig. 11-a), as the amount of non-recirculated sludge increases, organic matter concentration in the effluent increases, worsening effluent quality. For TSS (Fig. 11-b) the opposite is evident, as the sludge amount that is not recirculated increases, solids concentration decreases.

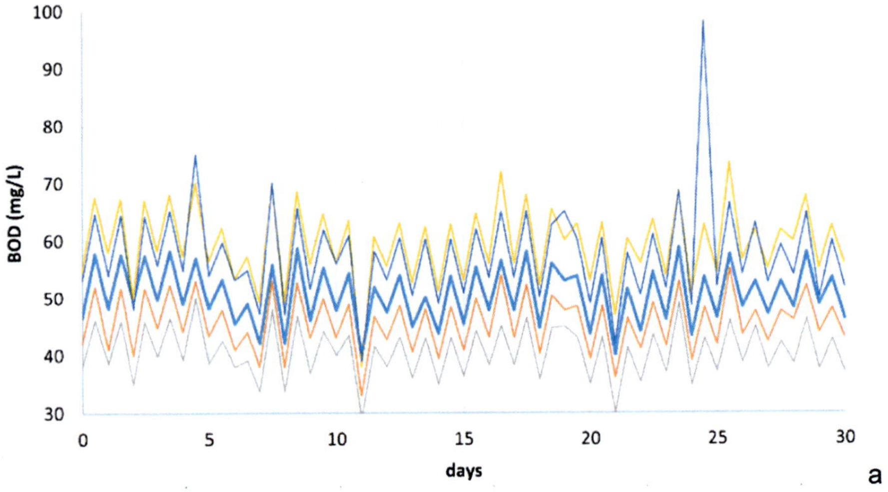

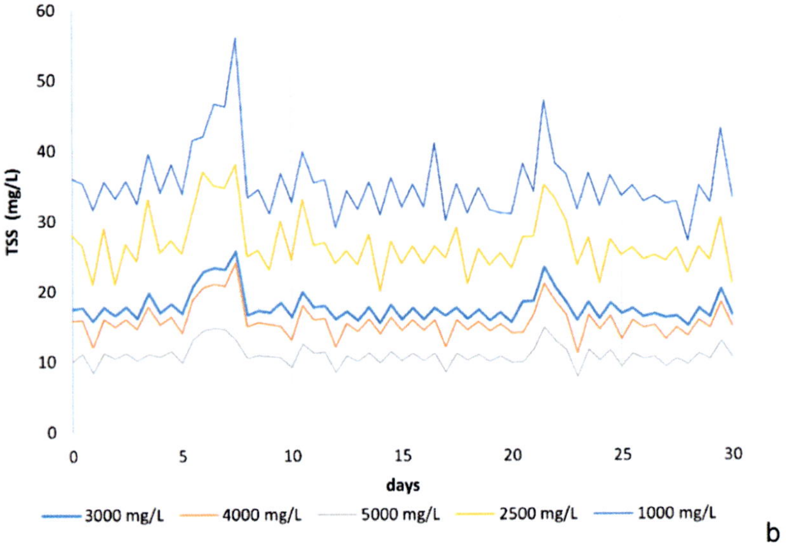

Fig. 11. a) BOD (mg/L) effluent and b) TSS (mg/L) effluent, depending on the internal recirculation of sludge.

From sensitivity analyzes carried out, it is concluded that operating parameters that improve the effluent quality in the greatest proportion are 2.5 mg/L and 1000 mg/L for dissolved oxygen and sludge recycling, respectively.

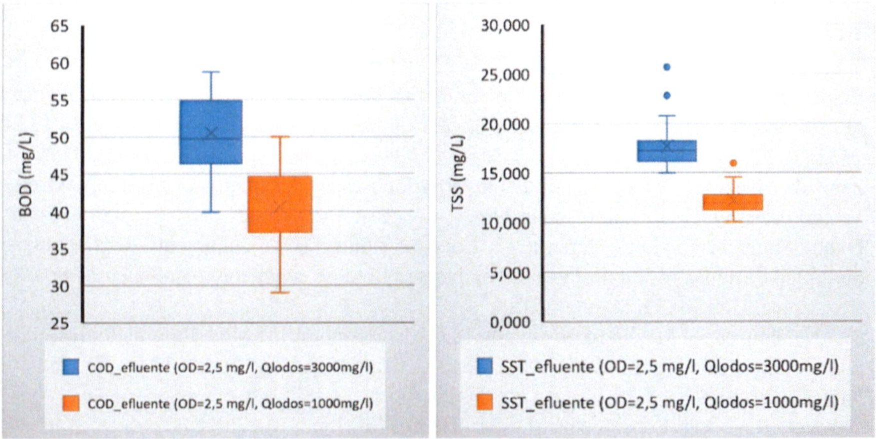

Fig. 12. Result of the simulation in BOD and SST: simulation with initial values (blue) vs. simulation with the optimal choice of DO and Qlodos (orange).

Finally, an assay was carried out with these concentrations to compare the effluent quality respect to initial operating conditions. In Fig. 12 it is observed that quality improves in terms of organic matter and suspended solids concentration, however, it will be necessary to continue evaluating other variables that could be conditioning factors regarding its operation and use, and how they affect the effluent performance and quality. This will be a future task in the present investigation, in addition, kinetic coefficients will also be evaluated, which for this case, were taken from those that are established by default in the WEST software; for this, subsequent tests to characterize residual water using respirometry techniques will be necessary.

4 Conclusions

Through the simulation with WEST, it was possible to obtain an approximation to the general behavior of the G-WWTP and consequently, a basic sensitivity analysis was developed on variables that can be manipulated at macro level in the plant: dissolved oxygen concentration and internal sludge recirculation. By comparing all these parameters, it can be concluded that dissolved oxygen variation other than 2.5 mg/L does not represent significant variations on the water effluent quality, on the other hand, a load of recirculated sludge is a parameter that improves the water quality significantly at a concentration of 1000 mg/L.

In this first approximation, the present research did not perform a calibration of the implemented model, consequently, the result simulations are approximations obtained from quantitative implemented models that were coupled to the available information. For this reason, based on inflow and effluent time series, it is recommended to perform a calibration on the implemented model to be able to validate the results obtained. In addition, carry out an exhaustive determination of kinetic coefficients for this type of wastewater. It motivates a future work of essays, tests, and complete modeling.

References

1. Bagley, D.M., Brodkorb, T.S.: Modeling microbial kinetics in an anaerobic sequencing batch reactor - model development and experimental validation. Water Environ. Res. **71**(7), 1320–1332 (1999)
2. Zec, M., Hvala, N.: Modelling and simulation of wastewater treatment plants. Elektroteh. Vestnik/Electrotech. Rev. **63**(4–5) (1996)
3. Duque Sarango, P., Heras Naranjo, C., Lojano Criollo, D.: Modelamiento del tratamiento biológico de aguas residuales; estudio en planta piloto de contactores biológicos rotatorios. Rev. Cienc. UNEMI **11**, 88–96 (2018)
4. Van Veldhuizen, H.M., Van Loosdrecht, M.C.M., Heijnen, J.J.: Modelling biological phosphorus and nitrogen removal in a full scale activated sludge process. Water Res. **33**(16), 3459–3468 (1999)
5. Gernaey, K.V., Van Loosdrecht, M.C.M., Henze, M., Lind, M., Jørgensen, S.B.: Activated sludge wastewater treatment plant modelling and simulation: state of the art. Environ. Model. Softw. **19**(9), 763–783 (2004)
6. Jerves-Cobo, R., et al.: Integrated ecological modelling for evidence-based determination of water management interventions in urbanized river basins: case study in the Cuenca River basin (Ecuador). Sci. Total Environ. **709**, 136067 (2020)
7. Gao, F., Nan, J., Li, S., Wang, Y.: Modeling and simulation of a biological process for treating different COD: N ratio wastewater using an extended ASM1 model. Chem. Eng. J. **332**, 671–681 (2018)
8. Duque-sarango, P., Cando, A., Segarra, F., Zhingri, A.: Revista Amazónica Ciencia y Tecnología. Rev. Amaz. Cienc. y Tecnol. **7**(3), 172–183 (2018)
9. Vanhooren, H., Meirlaen, J., Amerlinck, Y., Claeys, F., Vangheluwe, H., Vanrolleghem, P.A.: WEST: modelling biological wastewater treatment. J. Hydroinformatics **5**(1), 27–50 (2003)
10. Alvarado, A., Vedantam, S., Goethals, P., Nopens, I.: A compartmental model to describe hydraulics in a full-scale waste stabilization pond. Water Res. **46**(2), 521–530 (2012)
11. Greeley and Hansen. ACSAM Consultores: Estudio De Impacto Ambiental Para La Planta De Tratamiento De Aguas Residuales Guangarcucho, p. 317 (2017)
12. Song, Z., Williams, C.J., Edyvean, R.G.J.: Sedimentation of tannery wastewater. Water Res. **34**(7), 2171–2176 (2000)
13. Petersen, B., Gernaey, K., Henze, M., Vanrolleghem, P.A.: Calibration of activated sludge models: a critical review of experimental designs. In: Agathos, S.N., Reineke, W. (eds.) Biotechnology for the Environment: Wastewater Treatment and Modeling, Waste Gas Handling, pp. 101–186. Springer, Dordrecht (2003). https://doi.org/10.1007/978-94-017-0932-3_5
14. Altaher, H., Alghamdi, A.: Enhancement of quality of secondary industrial wastewater effluent by coagulation process: a case study. J. Environ. Prot. **02**(09), 1250–1256 (2011)
15. Asensi, E., Alemany, E., Duque-Sarango, P., Aguado, D.: Assessment and modelling of the effect of precipitated ferric chloride addition on the activated sludge settling properties. Chem. Eng. Res. Des. **150**, 14–25 (2019)

Variational Autoencoders for Assessing Sustainability

José Fernando Romero-Cañizares(✉) and Purificación Vicente-Galindo

Department of Statistics, University of Salamanca, Salamanca, Spain
{joseromero,purivg}@usal.es

Abstract. Reliable, impartial, and timely data models are urgently needed to inform global sustainability measures. As a response, implicit and explicit fuzzy methods that evaluate qualitative and quantitative variables have been proposed. Meanwhile, decision-makers are often compelled to address immediate over long-term risks. In the era of big data, artificial intelligence, and the internet we need to leverage the power of statistical models to support effective leadership. This article introduces a new technique for evaluating sustainability based on the Sustainability Assessment by Fuzzy Evaluation (SAFE) model: Variational Autoencoder plus graphical analysis (VAE&GA). This approach produces SAFE-like rankings backed by a dual axis chart that groups countries according to their most important unique indicators. VAE&GA is thus a more objective alternative to current fuzzy methods.

Keywords: Artificial neural network · Sustainability indicators · Graphical sustainability ranking

1 Introduction

Agenda 21, the main outcome of the 1992 Rio Earth Summit, stressed the need to develop indicators of sustainability. The Rio Summit of 2012 marked additional progress towards sustainability [1]. The 2000 Millennium Summit set the Millennium Development Goals (MDGs) to be achieved by the year 2015. On September 25th of that year, 193 member states of the United Nations adopted a set of 17 Sustainable Development Goals (SDGs) [2].

Machine learning uses artificial neural networks. An artificial neural network (ANN) is a computer system that is made up of a collection of connected units called neurons that are organized into layers. Neurons are commonly called nodes. Each connection between two nodes has an associated weight, which is a number that represents the strength of the connection between the two nodes. Layers positioned between the input and output layers are known as hidden layers [3].

A commonly used activation function is the rectified linear unit (ReLU), which transforms the input to the maximum of either 0 or the input itself. The more positive the neuron is, the more activated it is.

$$\mathrm{ReLU}\,(x) = \max\,(0,\; x) \tag{1}$$

S. Berrezueta and K. Abad (Eds.): *Doctoral Symposium on Information and Communication Technologies - DSICT*, LNEE 846, pp. 47–62, 2022.
https://doi.org/10.1007/978-3-030-93718-8_5

The loss function measures how similar the reconstructed version is to the original version. The more similar the reconstructed output is to the original input, the lower the loss. One common loss function is mean squared error (MSE). After analyzing our samples, the data is passed through our model repeatedly. This process of repeatedly sending the same data over the network is considered training. It is during this training process that the model will learn [3].

The weights are optimized using an optimization algorithm. The optimization process depends on the chosen optimization algorithm (optimizer). The best-known optimizer is called stochastic gradient descent (SGD).

During training when the loss is calculated for our inputs, the gradient of that loss is calculated for each of the weights in our model. The gradients are then multiplied by the learning rate.

$$\text{old weight} - (\text{learning rate} * \text{gradient})$$

This learning rate is a small number that typically ranges from 0.01 to 0.0001, but the actual value can vary, and whatever value is obtained for the gradient will become quite small once we multiply it by the learning rate [3].

To calculate the weight's new value, we use this formula:

$$\text{new weight} = \text{old weight} - (\text{learning rate} * \text{gradient}) \tag{2}$$

The principal machine learning scenarios are supervised, semi-supervised, and unsupervised learning. Supervised learning uses a series of labeled samples as training data to predict all unseen instances, which occur in problems of classification, regression, and ranking. This type of learning is applied in several areas such as anomaly detection, face recognition, signal and image classification, weather forecasting, and stock market forecasting [4]. In the case of semi-supervised learning, the learner receives a training samples, both labeled and unlabeled, and predicts all unseen instances. This is preferred because unlabeled sample are less costly than labelled samples. Classification, regression, and ranking processes are typical applications [5–7]. In unsupervised learning, the learner solely receives unlabeled samples from which training data is obtained and used to predict all unseen instances. Clustering and reduction of dimensionality [8, 9], as well as anomaly detection, and association rule mining [10] are some applications.

2 Principles of the Variational Autoencoder

A variational autoencoder is an autoencoder, used principally in unsupervised deep learning scenarios, whose encoding distribution is regularized during the training to achieve organized latent space representations (continuity and completeness), and use these representations to generate additional data. A variational autoencoder uses a latent vector model to analyze input data. Therefore, instead of letting the neural network learn an arbitrary function, it learns the parameters of a probability distribution that models the data [3].

First, an encoder network turns the input sample x into two parameters using a latent vector, called z_mean and z_log_sigma. Then, similar points are randomly sampled:

z from the latent normal distribution that is supposed to generate the data, via z = z_mean + exp(z_log_sigma) * epsilon, where epsilon is a random normal tensor. Finally, a decoder network maps these latent vector points back to the original input data (ibid).

The parameters of the model are trained via two loss functions: A **reconstruction loss** that forces the decoded samples to match the initial inputs, and the **Kullback-Leibler divergence** between the learned latent distribution and the prior distribution, acting as a **regularization term**. This helps to learn well-formed latent vectors and to reduce overfitting of training data [3].

The goal in unsupervised learning is to discover uniformities in the input data because particular patterns appear more frequently than others in the input space configuration [10].

Clustering techniques are machine learning approaches used to find similar patterns and relationships among data samples, and then cluster these samples into groups. Each cluster has some attributes or correlation characteristics. Because, they are fully unsupervised, they seek to cluster the samples without prior training, guidance, or knowledge of characteristics, relationships, and associations of the data [11].

K-means stands for learning predictive clustering, and we can learn a clustering model from training data. Therefore, considering that our database is numerical, we decided to use the Euclidean distance to minimize some distance-related quantity over all instances of the set. In addition, it is important to mention that in a K-means problem there is no effective solution to identify the global minimum. K-means converge in finite time to a stationary point; there can be no certainty whether the convergence point is the global minimum or not, no matter how far we are from it [12].

We propose using the variational autoencoder plus graphical analysis (VAE&GA) based on the Gaussian-distributed class centroids of latent variables to train the network to ensure the maximization of inter-class distance and the minimization of inner-class distance in assessing world sustainability.

3 Mathematical Remarks of Variational Autoencoders

We start from the initial SAFE data training set. Each country in the dataset has 69 sustainability indices. Therefore, the input layer has 69 nodes corresponding to each index. Then we encoded them in a latent space called z, generally of few dimensions (two in our case), and from this latent space, we reconstruct the initial data set with the greatest possible precision (Fig. 1).

The SAFE training data is represented by x, and it is assumed that it is generated from a latent variable z (the encoded representation) that is not directly observed. Thus, for each data point, a latent representation z is sampled from the prior distribution p(z), and the data x is sampled from the conditional likelihood distribution p(x|z) [13].

The "probabilistic decoder" defined by p(x|z) describes the distribution of the decoded variable given the encoded one, whereas the "probabilistic encoder" defined by p(z|x), describes the distribution of the encoded variable given the decoded one (ibid).

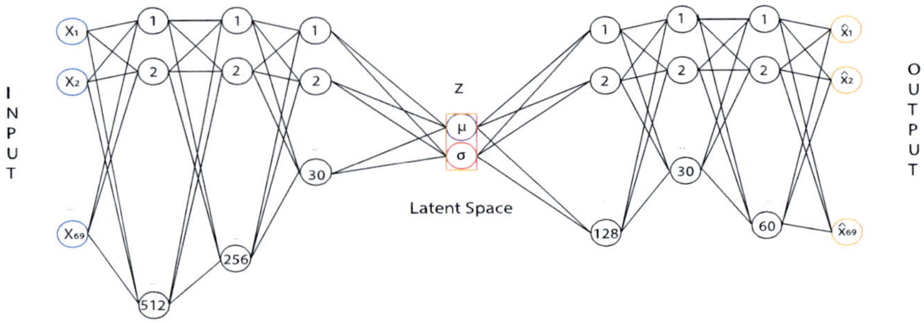

Fig. 1. Variational autoencoder graphical model

The link between the prior p(z), the likelihood p(x|z), and the posterior p(z|x) follows Bayes theorem:

$$p(z|x) = \frac{p(x|z)p(z)}{p(x)} = \frac{p(x|z)p(z)}{\int p(x|\mu)p(\mu)d\mu} \tag{3}$$

It is assumed that p(z) is a standard Gaussian distribution, and that p(x|z) is a Gaussian distribution whose mean is defined by a deterministic function f of the variable z and whose covariance matrix has the form of a positive constant c that multiplies the identity matrix I. The function f is assumed to belong to a family of functions denoted F.

$$\begin{aligned} &p(z) \equiv N(0, I) \\ &p(x|z) \equiv N(f(z), cI) \qquad f \in F \qquad c > 0 \end{aligned}$$

3.1 Variational Inference Formulation

The training of all neural networks is an iterative process where the error of the output data with some expected value is calculated, and this error is minimized in the following iterations. In this case, the output value $\hat{x}$ (data generated) is used and the error is calculated considering the original input data. In addition, the Kullback-Leibler (KL) divergence between z and $\hat{x}$ indicates how similar these two are [2].

We approximate p(z|x) with a Gaussian distribution q_x(z) whose mean and covariance are defined by two functions, g, and h, of the parameter x. These two functions belong to the families of the functions G and H respectively (ibid).

$$q_x(z) \equiv N(g(x), h(x)) \qquad g \in G \qquad h \in H$$

To find the best approximation among these families, it is necessary to minimize the Kullback-Leibler (KL) divergence between the approximation and the target p(z|x) to optimize the functions g and h. Thus, we are looking for the optimal g^* and h^* such that

$$\left(g^*, h^*\right) = \underset{(g, h) \in G \times H}{\arg\min} KL(qx(z), p(z|x)) \tag{4}$$

Replacing Bayes theorem such that

$$\left(g^{*},h^{*}\right) = \arg\min(\mathbb{E}\, z \sim qx(\log qx(z) - \mathbb{E}\, z \sim qx(\log \frac{P(x|z)P(z)}{p(x)}))$$
$$(g,h) \in GxH \tag{5}$$

Applying the properties of logarithms such that

$$= \arg\min(\mathbb{E}\, z \sim qx(\log qx(z)) - \mathbb{E}\, z \sim qx(\log p(z)) - \mathbb{E}\, z \sim qx(\log p(x|z)) + \mathbb{E}\, z \sim qx(\log p(x)))$$
$$(g,h) \in GxH \tag{6}$$

Maximizing with negative KL (-KLD)

$$= \arg\max(\mathbb{E}\, z \sim qx\,(\log p(x|z)) - KL(qx\,(z), p(z)))$$
$$(g,h) \in GxH \tag{7}$$

$$= \arg\max\,(\mathbb{E}\, z \sim qx\,(-\frac{\|x-f(z)\|^{2}}{2c}) - KL\,(qx(z), p(z)))$$
$$(g,h) \in GxH \tag{8}$$

In the penultimate Eq. (7), we can observe the tradeoff that exists approximating the posterior p(z|x), between maximizing the likelihood of the "observations" (maximization of the expected log-likelihood, for the first term) and staying close to the prior distribution (minimization of the KL divergence between q_x(z) and p(z), for the second term) [13].

Since p(z|x) can be approximated (by variational inference) from p(z) and p(z|x); and since p(z) is a simple standard Gaussian, the only two levers we have at our disposal in our model to make optimizations are the parameter c (which defines the variance of the probability) and the function f (which defines the mean of the probability) [13].

Consider that we can obtain for any function f in F (each defining a different probabilistic decoder p(x|z)) the best approximation of p(x|z)), denoted q* _x (z). Despite its probabilistic nature, we look for the most efficient encoding-decoding scheme possible, and then we choose the function f that maximizes the expected logarithmic probability of x given z when z is sampled from q*_x (z).

It is equivalent to saying that, for a given input x, we want to maximize the probability of having $\hat{x} = x$ when we sample z from the distribution q*_x (z) and then we sample $\hat{x}$ from the distribution p(x|z). Therefore, we look for the optimal f* such that:

$$f^{*} = \underset{f \in F}{\arg\max}\ \mathbb{E}\, z \sim q_{x}^{*}\,(\log p(x|z)) \tag{9}$$

$$= \underset{f \in F}{\arg\max}\ \mathbb{E}\, z \sim q_{x}^{*}(-\frac{||x-f(z)||^{2}}{2c}) \tag{10}$$

where q*_x(z) depends on the function f. Then, the optimal f*, g*, and h* is:

$$(f^{*},g^{*},h^{*}) = \arg\max(\mathbb{E}\, z \sim q_{x}(-\frac{||x-f(z)||^{2}}{2c}) - KL(qx(z), p(z)))$$

$$(f, g, h) \in FxGxH \tag{11}$$

This identifies the reconstruction error between x and f(z), the regularization term given by the KL divergence between q_x (z) and p(z) (which is a standard), and the constant c that governs the balance between the two previous terms. The larger the c is, the more we assume a high variance around f(z) for the probabilistic decoder in our model and, therefore, the more we favor the regularization term over the reconstruction term (the opposite occurs when c is low) [13].

3.2 Neural Networks into the Model

F, G, and H correspond to the families of functions defined by the network architecture and the optimization is carried out on the parameters of these networks.

In practice, g and h are not defined by two completely independent networks, but rather share a part of their architecture and their weights so that:

$$g(x) = g_2(g_1(x)) \qquad h(x) = h_2(h_1(x)) \qquad g_1(x) = h_1(x) \tag{12}$$

The covariance matrix of q_x (z), h(x) is a square matrix. To simplify the calculation and reduce the number of parameters, an additional assumption is made such that the approximation of p(z|x), q_x(z), is a multidimensional Gaussian distribution with diagonal covariance matrix (assumption of independence of variables). With this assumption, h(x) is simply the vector of the diagonal elements of the covariance matrix and therefore has the same size as g(x). However, this reduces the family of distributions considered for the variational inference and as a result, the approximation of p(z|x) obtained may be less precise (Fig. 2) [13].

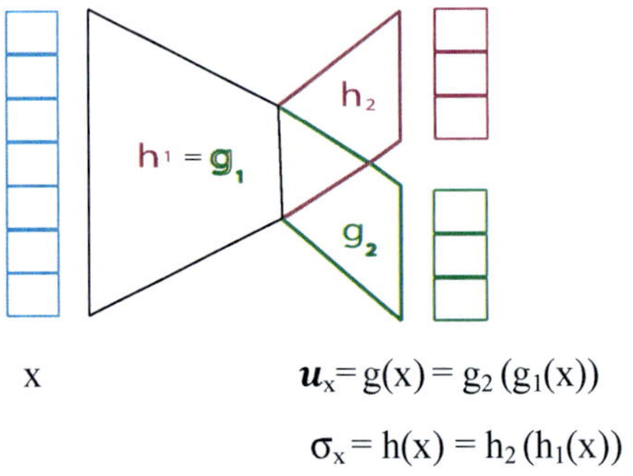

Fig. 2. Encoder part of the VAE [13].

Contrary to the part of the encoder that models p(z|x) and for which a Gaussian was considered with mean and covariance that are functions of x (g and h), the model assumes for p(z|x) a Gaussian with fixed covariance. The function f of the variable z that defines the mean of that Gaussian is modeled by a neural network and can be represented as follows (Fig. 3) [13]:

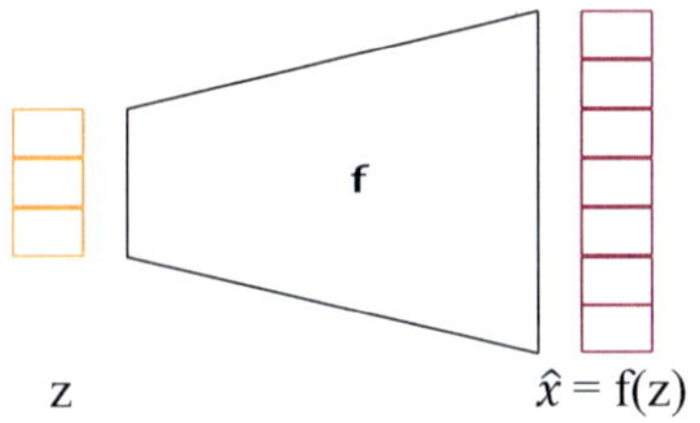

Fig. 3. Decoder part of the VAE [13].

The general architecture is then obtained by concatenating the encoder and decoder parts. However, we must be very careful about how we sample from the distribution returned by the encoder during training. The sampling process must be expressed in a way that allows the error to be backpropagated through the network. So, if z is a random variable that follows a Gaussian distribution with mean g(x) and with covariance H(x) = h(x).h ^ t(x), then it can be expressed as [13]:

$$z = h(x)\zeta + g(x) \qquad \zeta \sim N\,(0, I) \tag{13}$$

Finally, the objective function of the variational autoencoder architecture (Fig. 4) is given by the Eq. (11), in which the theoretical expectation is replaced by Monte-Carlo approximation. Then, considering this approximation and denoting C = 1/(2c), the loss function is recovered and is composed of a reconstruction term, a regularization term, and a constant to define the relative weights of these two terms [13].

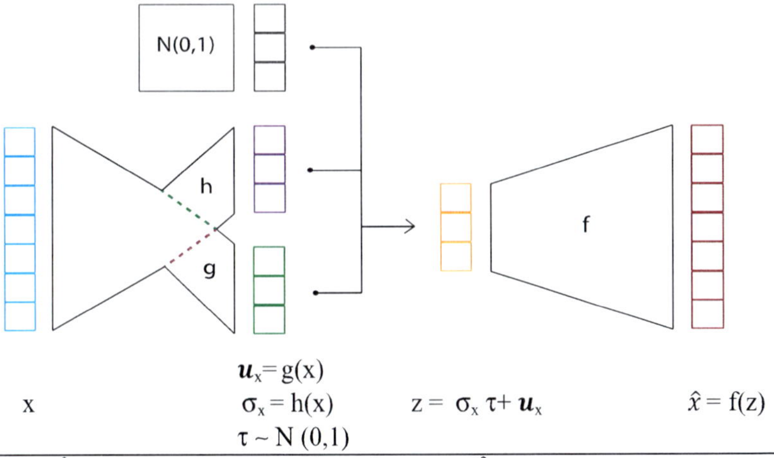

$$\text{Loss}=C\|x-\hat{x}\|^2+KL[N(u_x,\Sigma_x),N(0,1)] = C\|x-f(z)\|^2+KL[N(g(x),h(x)), N(0,1)] \tag{14}$$

Fig. 4. Variational autoencoders representation [13]

4 Outline of SAFE

Fuzzy logic is appropriate to deal with qualitative, imprecise, or uncertain information [14]. SAFE is a model that assesses sustainability using fuzzy logic. A system of fuzzy reasoning evaluates the various composite components of sustainability and sustainability via "if-then" rules. Finally, the output of the system is converted into a crisp value of sustainability utilizing a defuzzification process [15]. The model provides country rankings and performs sensitivity analyses that reveal key indices that each country should focus on to improve sustainability [15–17].

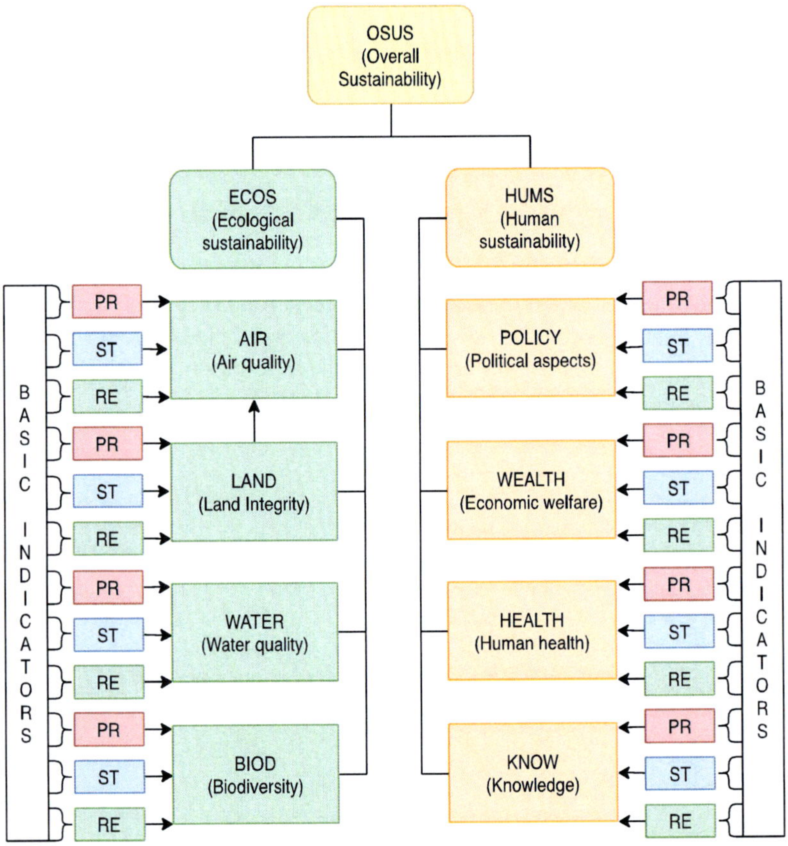

Fig. 5. The SAFE model

Basic indicator data are passed through an exponential smoothing filter to account for the memory of past performance and then are normalized on [0, 1] according to their sustainability standing, where 0 corresponds to unsustainable and 1 to totally sustainable values. Missing data are generated via an imputation procedure. Next, a multistage fuzzy inference engine is used together with pertinent rule bases to obtain fuzzy values for composite sustainability variables. A height defuzzification procedure yields crisp sustainability numbers at each stage. The final number of overall sustainability is used to rank countries. Finally, a sensitivity analysis reveals those indicators that have the greatest potential of improving sustainability.

Several basic indicators are used to compute the four components of the ecosystem dimension: air, land, water, and biodiversity, and the four components of the human dimension: policies, wealth, health, and knowledge. Finally, an index of overall sustainability (OSUS) in [0, 1] is derived for each country. The hierarchical structure of the indicators in the SAFE model is outlined in Fig. 5.

5 Results and Discussion

Unsupervised techniques attempt to learn fundamental latent structures, relations, and patterns of information without any assistance or supervision [10], and after defining the parameters and optimizing the membership functions, the process is very time efficient because it is based on the training data rather than the expert knowledge [18]. After training and checking, there is a 73% coincidence (120 out of 164 countries) in the country grouping memberships (Table 1).

To evaluate the sustainability of countries, we divide sustainability into 3 main clusters using K-means, as follows: cluster 1 (developed countries), cluster 2 (intermediate countries), and cluster 3 (developing countries). To complete the graphical analysis, we forced sustainability indicators to fall on two axes: X and Y, and we have a good picture of sustainability indicators and country groups (Fig. 6).

The most sustainable countries were placed in the lower left quadrant of Fig. 6 and form group 1 (Table 1). The most influential indicators on the negative X axis belong to both the environmental and human dimensions, and on the negative Y axis, all of them belong to the human dimension (Table 2).

The least sustainable countries were placed in the upper right quadrant of Fig. 6 and form group 3 (Table 1). The most influential indicators are shown along the positive X axis, all of them belong to the environmental dimension, and on the positive Y axis, they belong to both the human and environmental dimensions (Table 2).

Although VAE&GA is not a hierarchical technique, differences between SAFE and VAE are shown in Table 3.

Table 1. Clustering membership (VAE&GA) for 164 countries worldwide (SAFE 2020)

	Country	Cl		Country	Cl		Country	Cl		Country	Cl
1	Denmark	1	42	Russia	2	83	Bosnia and Herz	1	124	Rwanda	3
2	Sweden	1	43	Brunei	2	84	Zimbabwe	3	125	South Africa	3
3	Norway	1	44	Albania	1	85	Armenia	1	126	Cameroon	3
4	Switzerland	1	45	Cuba	1	86	Honduras	2	127	Mali	3

(*continued*)

Table 1. (*continued*)

	Country	Cl		Country	Cl		Country	Cl		Country	Cl
5	Austria	1	46	Peru	2	87	Seychelles	2	128	Swaziland	3
6	Finland	1	47	Brazil	2	88	Colombia	2	129	Belize	2
7	Slovenia	1	48	Morocco	2	89	El Salvador	2	130	Un. Arab Emirates	2
8	Netherlands	1	49	Mexico	2	90	Vietnam	3	131	Papua N.G	3
9	Slovakia	1	50	Moldova	1	91	Botswana	2	132	Bahamas	2
10	United Kingdom	1	51	Venezuela	2	92	Nepal	3	133	Liberia	3
11	France	1	52	Thailand	2	93	Eq. Guinea	3	134	Ethiopia	3
12	Lithuania	1	53	Argentina	2	94	Lao PDR	2	135	Egypt	2
13	Iceland	1	54	Mongolia	2	95	Burkina Faso	3	136	Bahrain	2
14	Germany	1	55	Azerbaijan	2	96	Kiribati	2	137	Nigeria	3
15	Poland	1	56	Tunisia	2	97	Congo Rep	3	138	Tanzania	3
16	Hungary	1	57	Serbia	1	98	Tajikistan	2	139	Djibouti	3
17	Ireland	1	58	Ghana	3	99	Jordan	2	140	Qatar	2
18	Czech Rep	1	59	Paraguay	2	100	Algeria	2	141	Angola	3
19	Portugal	1	60	North Macedonia	1	101	Senegal	3	142	Mozambique	3
20	Latvia	1	61	Kyrgyzstan	1	102	Iran	2	143	Madagascar	3
21	Estonia	1	62	Malaysia	2	103	Saudi Arabia	2	144	Guinea	3
22	Spain	1	63	Namibia	3	104	Indonesia	2	145	Gambia	3
23	Croatia	1	64	Gabon	3	105	Togo	3	146	Oman	2
24	Australia	1	65	Israel	2	106	Kenya	3	147	Myanmar	3
25	Italy	1	66	Singapore	2	107	Turkmenistan	2	148	Burundi	3
26	Belgium	1	67	Dominican Rep	2	108	Maldives	2	149	Uganda	3
27	Malta	1	68	Ukraine	1	109	China	2	150	Bangladesh	2
28	Uruguay	2	69	Cape Verde	2	110	Malawi	3	151	Guatemala	3
29	Luxembourg	1	70	Fiji	2	111	Trinidad and Tobago	2	152	Pakistan	3
30	New Zealand	1	71	South Korea	2	112	Uzbekistan	2	153	Chad	3

(*continued*)

Table 1. (*continued*)

	Country	Cl		Country	Cl		Country	Cl		Country	Cl
31	Canada	1	72	Ecuador	2	113	Côte d'Ivoire	3	154	Niger	3
32	Greece	1	73	Georgia	1	114	Lesotho	3	155	Congo D.R	3
33	Cyprus	1	74	Panama	2	115	Benin	3	156	Iraq	3
34	Romania	1	75	Guyana	2	116	Sierra Leone	3	157	Guinea-Bissau	3
35	Japan	1	76	Nicaragua	2	117	Cambodia	3	158	Yemen	3
36	United States	1	77	Philippines	2	118	Jamaica	2	159	Central African Rep	3
37	Bulgaria	1	78	Suriname	2	119	Zambia	3	160	Eritrea	3
38	Costa Rica	2	79	Kazakhstan	1	120	Kuwait	2	161	Sudan	3
39	Belarus	1	80	Bolivia	2	121	Libya	2	162	Mauritania	3
40	Chile	2	81	Mauritius	2	122	Lebanon	2	163	Haiti	3
41	Turkey	2	82	Sri Lanka	2	123	India	3	164	Afghanistan	3

Table 2. Contribution of variables to each axis (SAFE indicators 2020)

	Positive	Negative
Factorial Axis 1	EL10-Fertilizers EA1-CO2_emissions EA3-NOx_emissions EW19-Water_stress	EB24-Fish_stock_status HH4-Mortality_PM HP37-Inflation EW20-Water_resources HW42-Foreing direct investment
Factorial Axis 2	HH43-Cardiovascular_incidence EL18-Hazardous_w HH44-Neoplastic_incidence EL9-Pesticides EA2-SO2_emissions HP39-Government_debt EL11-SLR_land_impact EL14-Forest_area (trend) HP36-Military_expenditure EL18-Terrestrial_prot_area EW35-Unemployment_gender_inequality	HH47-Malaria_incidence HH54-Physicians HW40-GNI HP31-Civil_ liberties HP47-Political_ rights HH55-Hospital_beds HH58-Access_sanitation HK65-Secondary_school_enrollement HH48-Infant_mortality HP33-Corruption HH49-Maternal_mortality

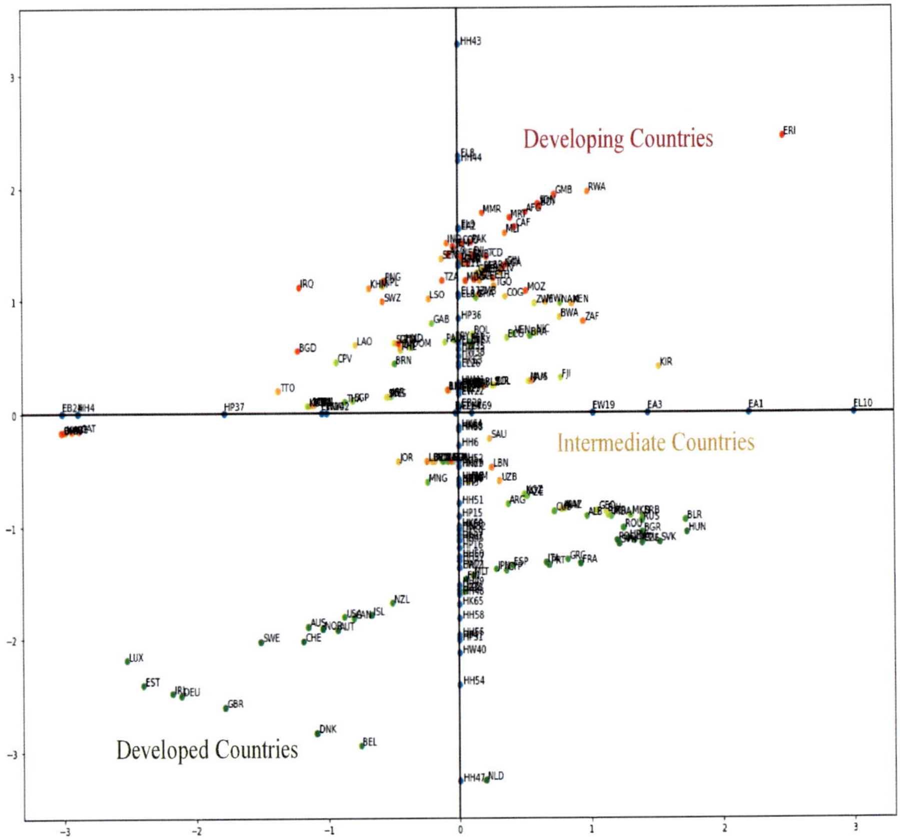

Fig. 6. VAE&GA worldwide country sustainability performance

Table 3. Ranking of the differences between SAFE and VAE for 164 countries worldwide.

Country	Rank	VAE	(−)	Country	Rank	VAE	(−)
Denmark	1	3	−2	Russia	42	38	4
Sweden	2	9	−7	Brunei	43	94	−51
Norway	3	12	−9	Albania	44	41	3
Switzerland	4	10	−6	Cuba	45	46	−1
Austria	5	11	−6	Peru	46	105	−59
Finland	6	19	−13	Brazil	47	109	−62
Slovenia	7	28	−21	Morocco	48	63	−15
Netherlands	8	1	7	Mexico	49	101	−52
Slovakia	9	30	−21	Moldova	50	40	10
United Kingdom	10	4	6	Venezuela	51	111	−60

(continued)

Table 3. (*continued*)

Country	Rank	VAE	(−)	Country	Rank	VAE	(−)
France	11	25	−14	Thailand	52	76	−24
Lithuania	12	31	−19	Argentina	53	50	3
Iceland	13	16	−3	Mongolia	54	53	1
Germany	14	5	9	Azerbaijan	55	51	4
Poland	15	33	−18	Tunisia	56	65	−9
Hungary	16	35	−19	Serbia	57	42	15
Ireland	17	6	11	Ghana	58	122	−64
Czech Rep	18	29	−11	Paraguay	59	107	−48
Portugal	19	24	−5	North Macedonia	60	43	17
Latvia	20	32	−12	Kyrgyzstan	61	52	9
Estonia	21	7	14	Malaysia	62	79	−17
Spain	22	23	−1	Namibia	63	118	−55
Croatia	23	34	−11	Gabon	64	113	−49
Australia	24	13	11	Israel	65	80	−15
Italy	25	26	−1	Singapore	66	77	−11
Belgium	26	2	24	Dominican Rep	67	99	−32
Malta	27	20	7	Ukraine	68	44	24
Uruguay	28	18	10	Cape Verde	69	95	−26
Luxembourg	29	8	21	Fiji	70	92	−22
New Zealand	30	17	13	South Korea	71	72	−1
Canada	31	14	17	Ecuador	72	108	−36
Greece	32	27	5	Georgia	73	47	26
Cyprus	33	21	12	Panama	74	104	−30
Romania	34	37	−3	Guyana	75	86	−11
Japan	35	22	13	Nicaragua	76	112	−36
United States	36	15	21	Philippines	77	97	−20
Bulgaria	37	36	1	Suriname	78	88	−10
Costa Rica	38	84	−46	Kazakhstan	79	49	30
Belarus	39	39	0	Bolivia	80	110	−30
Chile	40	85	−45	Mauritius	81	90	−9
Turkey	41	62	−21	Sri Lanka	82	83	−1
Bosnia and Herz	83	45	38	Rwanda	124	163	−39
Zimbabwe	84	117	−33	South Africa	125	114	11

(*continued*)

Table 3. (*continued*)

Country	Rank	VAE	(−)	Country	Rank	VAE	(−)
Armenia	85	48	37	Cameroon	126	144	−18
Honduras	86	106	−20	Mali	127	155	−28
Seychelles	87	78	9	Swaziland	128	120	8
Colombia	88	89	−1	Belize	129	87	42
El Salvador	89	103	−14	Un. Arab Emirates	130	70	60
Vietnam	90	74	16	Papua N.G	131	130	1
Botswana	91	115	−24	Bahamas	132	82	50
Nepal	92	128	−36	Liberia	133	140	−7
Eq. Guinea	93	136	−43	Ethiopia	134	135	−1
Lao PDR	94	100	−6	Egypt	135	64	71
Burkina Faso	95	139	−44	Bahrain	136	68	68
Kiribati	96	93	3	Nigeria	137	141	−4
Congo Rep	97	123	−26	Tanzania	138	132	6
Tajikistan	98	54	44	Djibouti	139	149	−10
Jordan	99	58	41	Qatar	140	71	69
Algeria	100	60	40	Angola	141	150	−9
Senegal	101	145	−44	Mozambique	142	125	17
Iran	102	61	41	Madagascar	143	131	12
Saudi Arabia	103	66	37	Guinea	144	142	2
Indonesia	104	98	6	Gambia	145	162	−17
Togo	105	129	−24	Oman	146	67	79
Kenya	106	116	−10	Myanmar	147	158	−11
Turkmenistan	107	56	51	Burundi	148	160	−12
Maldives	108	73	35	Uganda	149	134	15
China	109	75	34	Bangladesh	150	96	54
Malawi	110	119	−9	Guatemala	151	102	49
Trinidad and Tobago	111	81	30	Pakistan	152	154	−2
Uzbekistan	112	55	57	Chad	153	148	5
Côte d'Ivoire	113	138	−25	Niger	154	147	7
Lesotho	114	121	−7	Congo D.R	155	152	3
Benin	115	137	−22	Iraq	156	127	29
Sierra Leone	116	133	−17	Guinea-Bissau	157	146	11
Cambodia	117	126	−9	Yemen	158	151	7
Jamaica	118	91	27	Central African Rep	159	156	3
Zambia	119	124	−5	Eritrea	160	164	−4

(*continued*)

Table 3. (*continued*)

Country	Rank	VAE	(−)	Country	Rank	VAE	(−)
Kuwait	120	69	51	Sudan	161	161	0
Libya	121	59	62	Mauritania	162	157	5
Lebanon	122	57	65	Haiti	163	143	20
India	123	153	−30	Afghanistan	164	159	5

6 Conclusions

Fuzzy logic has proven to be an uncontested numerical method as it occurs with SAFE.

An unsupervised learning method called Variational Autoencoder, and Graphical Analysis (VEA & GA) has proven effective in assessing sustainability using appropriate training data.

The promising results show that VAE&GA can be a sound alternative to assess sustainability, extrapolating its applications to other kinds of problems at different levels of analysis (continents, regions, cities, etc.) further corroborating the effectiveness of the unsupervised training methods.

References

1. Waas, T., Hugé, J., Verbruggen, A., Wright, T.: Sustainable development: a bird's eye view. Sustainability **3**(10), 1637–1661 (2011)
2. Sachs, J., Schmidt-Traub, G., Kroll, C., Durand-Declare, D., Teksoz, K.: SDG index and dashboards - global report. Bertelsmann Stiftung and Sustainable Development Solutions Network (SDSN), New York. https://www.sdgindex.org/reports/sdg-index-and-dashboards-2016/
3. Deeplizard: Machine Learning & Deep Learning Fundamentals, 21 November 2017
4. Zhu, Q., Zhang, R.: A classification supervised auto-encoder based on predefined evenly-distributed class centroids. arXiv:1902.00220v3 [cs.CV] (2020)
5. Fu, H., Lei, P., Tao, H., Zhao, L., Yang, J.: Improved semi-supervised autoencoder for deception detection. PLoS One **14**(10). https://doi.org/10.1371/journal.pone.0223361
6. Fu, X., et al.: Semi-supervised aspect-level sentiment classification model based on variational autoencoder. Knowl.-Based Syst. **171**, 81–92 (2019). https://doi.org/10.1016/j.knosys.2019.02.008
7. Zhu, Q., Li, T.: Semi-supervised learning method based on predefined evenly-distributed class centroids. Appl. Intell. **50**, 2770–2778 (2020). https://doi.org/10.1007/s10489-020-01689-1
8. Chorowski, J., Weiss, R.J., Bengio, S., van den Oord, A.: Unsupervised speech representation learning using WaveNet autoencoders. IEEE/ACM Trans. Audio Speech Lang. Process. **27**(12), 2041–2053 (2019). https://doi.org/10.1109/TASLP.2019.2938863
9. Yusiong, J.P.T., Naval, P.C.: AsiANet: autoencoders in autoencoder for unsupervised monocular depth estimation. In: Proceedings - 2019 IEEE Winter Conference on Applications of Computer Vision, WACV 2019, pp. 443–451 (2019). https://doi.org/10.1109/WACV.2019.00053
10. Bengio, Y.: Learning deep architectures for AI. Found. Trends Mach. Learn. **2**(1), 1–27 (2009). https://doi.org/10.1561/2200000006

11. Sarkar, D., Bali, R., Sharma, T.: Practical Machine Learning with Python. Apress, Berkeley (2018). https://doi.org/10.1007/978-1-4842-3207-1
12. Flach, P.: Machine Learning: The Art and Science of Algorithms that Make Sense of Data (2012). http://2.180.2.83:801/opac/temp/11802.pdf
13. Rocca, J.: Understanding Variational Autoencoders (VAEs). Building, step by step, the reasoning that leads to VAEs. Towards Data Science (2019). https://towardsdatascience.com/understanding-variational-autoencoders-vaes-f70510919f73
14. Zadeh, L.A.: Similarity relations and fuzzy orderings. Inf. Sci. **3**(2), 177–200 (1971)
15. Grigoroudis, E., Kouikoglou, V.S., Phillis, Y.A.: SAFE 2013: Sustainability of countries updated. Ecol. Ind. **38**, 61–66 (2014)
16. Grigoroudis, E., Kouikoglou, V., Phillis, Y.: SAFE 2019: Updates and new sustainability findings worldwide. Ecol. Ind. **121**, 107072 (2019)
17. Phillis, Y.A., Grigoroudis, E., Kouikoglou, V.S.: Sustainability ranking and improvement of countries. Ecol. Econ. **70**(3), 542–553 (2011)
18. Tan, Y., Shuai, C., Jiao, L., Shen, L.: Adaptive neuro-fuzzy inference system approach for urban sustainability assessment: a China case study. Sustain. Dev. **26**(6), 749–764 (2018). https://doi.org/10.1002/sd.1744/

Methods for Transforming Observable to Latent Variables of Adolescent Eating Behavior Using Mathematical Tools and Social Behavior Criteria

Julio Fernando Siguencia[1](✉), Mariela Cerrada[1], Diego Cabrera[1], René Vinicio Sánchez[1], and Angélica Ochoa[2]

[1] Universidad Politécnica Salesiana, Cuenca, Ecuador
jsiguenciau@est.ups.edu.ec
[2] Universidad de Cuenca, Cuenca, Ecuador

Abstract. Social factors such as intelligence, attitude, self-esteem (latent variables) are variables that provide relevant information to facilitate the generation of behavioral patterns or models for experts in the social area. However, such information cannot be measured directly, since they are not quantifiable. There are underlying characteristics (observable variables) to the social factors which can be measured directly to the subject of study. For this reason, a methodology based on the transformation of variables using mathematical tools is proposed. The proposed objective is to transform observable variables into hidden variables by three methods; using the arithmetic mean, Euclidean distance and exponential function. Finally, a metric based on EMD distances is applied to evaluate the similarity of the concept resulting from the three transformation methods. The EMD metric allows to evaluate the cost paid for taking one form of distribution to another, in this case the arithmetic mean and exponential function methods generate the lowest cost, that is, there is greater similarity between the distributions of the observable variables and the distribution of the resulting construct.

Keywords: EMD · Latent variable · Observable variable

1 Introduction

The data that provide social, psychological and emotional information, among others, play an important role in the study of human behavior and habits. These constructs allow the association of key behaviors to identify inappropriate behaviors of people [1, 2]. They are generally non-quantifiable data and are also known as constructs or latent variables, these characteristics make it possible to describe and explain the behavioral relationships regarding the lifestyle that people lead. A common problem faced by social researchers is to obtain or quantify these types of constructs, since they cannot be measured directly but through underlying indicators which are measurable through surveys, interviews and others. These underlying indicators are also known as visible or observable variables and can be obtained directly from the individual subject of the study. The perception of

S. Berrezueta and K. Abad (Eds.): *Doctoral Symposium on Information and Communication Technologies - DSICT*, LNEE 846, pp. 63–73, 2022.
https://doi.org/10.1007/978-3-030-93718-8_6

consumers is a clear example of a non-observable variable because it cannot be measured directly; however, in [3] an investigation is carried out on the relationship between the perception of consumers regarding nutritional information on beef where the data are the results of surveys with information on socioeconomic aspects, questions related to the perception of beef quality, costs and preferences of beef cuts and willingness to pay for a beef quality label. For instance, a study carried out in a community in southern Chile identified predictive factors associated with therapeutic non-compliance in hypertensive older adults. In this study, questionnaires are applied to measure levels of affection and to identify latent variables that could be formed from correlations of the scores of the different questionnaires. An exclusive analysis of observable variables does not allow us to describe the factors of therapeutic non-compliance because the information cannot be measured only in health facilities; it is necessary to know the primary environment in the home and to adapt health care based on the needs detected there in the population analyzed [4].

The quantitative methods in the social sciences use quantifiable information to generate knowledge about relationships between components, model behaviors about social reality or generally explain study phenomena [5]. This generated knowledge is part of the research process and the tools for the social scientific community to generate original contributions, confirmation or refutation of existing findings. An important issue to increase the knowledge to be generated is the number of study variables that can be collected; however, the greater the number of variables, the greater the complexity to process all this information. Due to technological advances, algorithm development and computational power, multivariate or multivariable methods have been implemented, which enable the simultaneous analysis of large sets of variables. This set of variables can be associated to behavioral factors and these in turn allow knowing social and psychological relationships, if applicable and according to the area of research. Nowadays, there are a number of multivariate methods and techniques for data analysis such as multiple regression, factor analysis, multivariate analysis of variance, causal analysis, structural equations, among others [6–10].

In [11] for example, a parametric and semiparametric latent variable transformation model approach is proposed to explore dysphagia, researchers to determine the effect of clinical and treatment factors on unobservable dysphagia. These models allow direct estimation of the treatment effect on the unobserved latent variable, which greatly improves the interpretability of the models. On the other hand, research work focused on worker health and safety in the mining industry uses latent variable modeling to improve understanding of the underlying constructs and their magnitude of effect versus observable parameters. The quantitative methods for latent variable transformation and analysis used in this study are confirmatory factor analysis (EFA), confirmatory factor analysis (CFA) and structural equation modeling with latent variables (SEM). Exploratory factor analysis studies the underlying relationships between variables by grouping them conceptually, based on an examination of appropriate statistics such as covariance or correlation [12]. The literature recommends the use of multivariate methods for the transformation of observable variables to latent variables because in most cases they generate excellent results; however, in some cases the data do not always follow a multivariate normal distribution and have different scales and weights.

This article presents three methods for transforming observable variables into latent variables, using statistical and mathematical tools such as the mean, euclidean distance and the natural exponential function. The criteria of belonging of observable variables to constructs are given by the experience of professionals related to the area of study, and the way to associate or transform a group of variables to a construct is done by statistical and mathematical tools that allow a smooth or adequate transformation process. From the statistical or mathematical point of view, the transformation cannot abruptly affect the distribution of the data, otherwise the information would be misrepresented.

2 Variables Transformation

The transformation of observable variables to latent variables is a vital part of the research process, since the reliability of the results depends on it. In this work, the transformation of variables is approached from a mathematical point of view to evaluate each method and assign a reliability criterion to each of them. In the Fig. 1 shows the methodology for transforming variables and is composed of observable variables as input to the system, methods for transforming variables and the resulting latent variable.

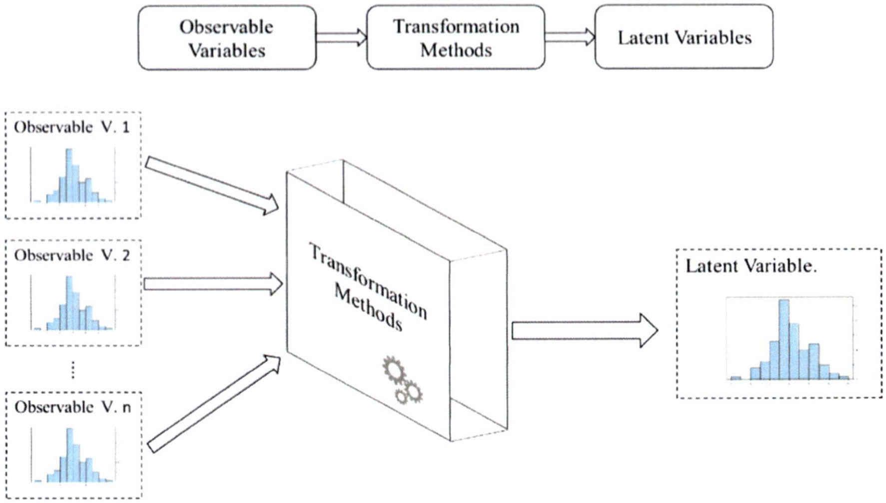

Fig. 1. Transformation process of observable variables to latent variables.

2.1 Observables Variables

Observable variables are a set of characteristics that can be measured directly on the individual being studied. Being social studies, the way to measure emotional, psychological and cultural behaviors, among others, is through interviews to the individuals under study, surveys (to the individual or third parties that surround him/her), anthropometric measurements and dietary components that can be associated, or behavioral constructs defined by experts in the area.

The following are questions related to the dietary behavior of young people and children in Ecuador. These questions address issues of self-efficacy, perceived benefits, parental permissiveness, barriers to healthy consumption, eating habits, among others. Each of these questions are formulated by professionals in nutrition and psychology so that the participants' responses are delimited between the categories: very difficult-very easy and totally disagree-totally agree, as shown in Table 1.

- How difficult is it for you to eat healthy every day?
- How difficult is it for you to eat healthy at home?
- How difficult is it for you to eat healthy at school?
- If I eat healthy, does it help me control my body weight?
- If I eat healthy, does it make me feel good?
- Do I eat snacks or fast food when I watch TV?

Table 1. Numerical assignment of categorical variables.

Self-efficacy	Perceived benefits, perceived barriers, habit strength
Very difficult: 1 Difficult: 2 Sometimes difficult, sometimes not difficult: 3 Easy: 4 Very easy: 5	Strongly disagree: 1 Disagree: 2 Neither agree nor disagree: 3 Agree: 4 Strongly agree: 5
Association with key components of dietary behavior	
Lowest intake: 1 Low intake: 2 Average intake: 3 Highest intake: 4 Highest intake: 5	

2.2 Transformation Method

Each construct is associated with a set of questions that were asked to the participants based on the criteria of the specialists in nutrition and psychology, however, it is necessary to apply a method of numerical transformation with the information acquired from the subjects. The quantifiable information as mentioned in the previous section allows the application of mathematical tools to generate relationships, behavioral models and others that facilitate the interpretation of the data. In this article, three methods of mathematical transformation of variables are proposed, such as: Application of the Arithmetic Mean, Application of Euclidean distance and Application of Exponential function. They are discussed in detail below.

2.2.1 Arithmetic Mean

In order to associate a certain number of questions to a concept, the arithmetic mean is used. In this case study, each of the applied questions has responses from the participants in the form of categorical data. The categorical data range from "Strongly agree" to "Strongly disagree". The categorical data are transformed into numerical data according to the categories by assigning a value of 5 to "Strongly agree" and a value of 1 to "Strongly disagree" to finally average the questions associated with each concept. To make the mean, each of the variables associated with a concept is added up and divided by the number of variables added, obtaining at the end a single vector of data as shown in Table 2.

Table 2. Transformation of variables using the arithmetic mean.

N°	Questions [n × m]				Construct [1 × n]
	QN. 1 [1–5]	QN. 2 [1–5]	...	QN. M [1–5]	(Mean) [1–5]
1	V1Q1	V1Q2	...	V1Qm	Sum(V1Q1 + V1Q2 + ... + V1Qm)/m
2	V2Q1	V2Q2	...	V2Qm	Sum(V2Q1 + V2Q2 + ... + V1Qm)/m
⋮	⋮	⋮	⋮	⋮	⋮
n	VnQ1	VnQ2	...	**VnQm**	Sum(VnQ1 + VnQ2 + ... + VnQm)/m

Therefore, **VnQm** represents the position of the data corresponding to the **n** participant and the **m** question applied.

2.2.2 Euclidean Distance

If a survey is correctly conducted, the questions are not correlated with each other, i.e. they are independent. Seen from a vector perspective, there is orthogonality between them. Assuming that this is true, the concept can be represented as a point in the n-dimensional space. Where the dimension is given by the number of questions associated with the concept and the answers to the questions represent the coordinates of the point in the vector space, as shown in Fig. 2.

To characterize the latent variable, the distance between the n-dimensional point and its origin can be taken. One of the metrics that allow to characterize these distances is the L2 norm or Euclidean norm. The Euclidean norm between two points in an N dimension is given by the Eq. 1:

$$d(A,B) = \sqrt{\sum_{i=1}^{N} (b_i - a_i)^2} = \sqrt{(b_1 - a_1)^2 + (b_2 - a_2)^2 + \ldots + (b_N - a_N)^2} \quad (1)$$

This equation allows to find the distance between two points in a vector space of N dimensions.

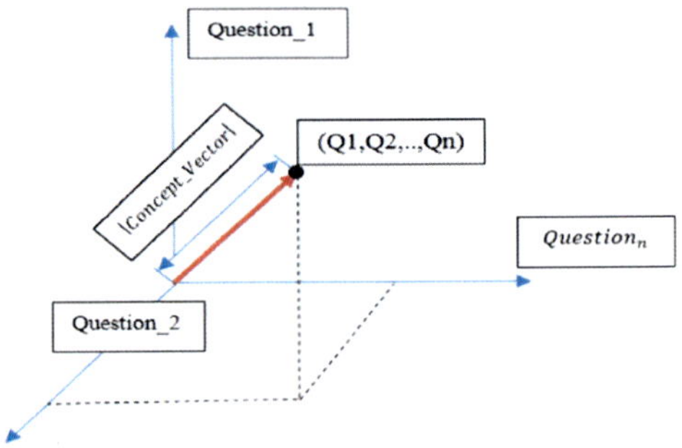

Fig. 2. Graphical representation of the vector analysis model.

2.2.3 Exponential Function

Assuming that the observational data are independent, this method seeks to assign a weight to the concept according to the response level of its associated questions, i.e., if the concept is associated with three questions of which the first two answers are assigned a value of 5 and the third a value of 1, the concept has a higher weight. The first idea is to weight the concept as the answers are permuted with the satisfiability values of the surveys conducted. The exponential function is used to assign a metric or weight to the concept.

The exponential function, as a technique for transforming observable variables to latent variables, begins to adjust each survey response as an independent variable of the exponential function. For obtaining the value of concept, a summation is performed on the dependent variable generated with the natural exponential function. This process is carried out with the total data obtained, respecting the association of questions to concepts given for specialists in the area studied. The result of applying the exponential function generates very high values, so it is necessary to smooth them to have a correct scaling and that is why the natural logarithm is applied to this result Finally, it is scaled according to the number of questions belonging to the concept. Each concept value is scaled separately since the concepts have a different number of variables assigned to them, as shown in Table 3.

2.3 Latent Variables

Latent variables cannot be measured directly, but can be inferred by mathematical modeling or through association with observable variables. These types of variables are also known as constructs that facilitate the understanding of human behavior. For example, the constructs and their underlying observable variables are shown below. It is worth mentioning that the association of questions to constructs is given by professionals in the area based on their experience.

- **Self-efficacy:** This concept is related to the personal capacity to consume healthy foods.
- **Perceived Benefits:** The questions in this concept try to find out if the child or adolescent is aware of the benefits of healthy eating.

Table 3. Transformation of variables using the exponential function.

N°	Questions [n × m]				Construct [1 × n]
	Qn. 1 [1–5]	Qn. 2 [1–5]	···	Qn. M [1–5]	(Mean) [1–5]
1	V1Q1	V1Q2	···	V1Qm	Ln[Sum(exp(V1Q1) + exp(V1Q2) + ... + exp(V1Qm))]
2	V2Q1	V2Q2	···	V2Qm	Ln[Sum(exp(V2Q1) + exp(V2Q2) + ... + exp(V2Qm))]
⋮	⋮	⋮	⋮	⋮	⋮
n	VnQ1	VnQ2	···	**VnQm**	Ln[Sum(exp(VnQ1) + exp(VnQ2) + ... + exp(VnQm))]

- **Perceived Barriers:** It is related to the barriers and behaviors that prevent healthy eating.
- **Habit Strength:** The child's ability to create a healthy eating habit.

In addition, environmental factors such as:

- **Accessibility Healthy Foods:** The concept relates to how accessible the child or adolescent is at home and at school to consume healthy foods.
- **Parental Permissiveness:** The permissiveness of parents to consume sugary drinks and fast food.
- **School Support:** Relates to questions that identify whether children receive guidance for healthy eating at school.
- **Association with key components of dietary behavior** (fruits, vegetables, sugary drinks, breakfast and consumption of unhealthy snacks).

3 Earth Mover's Distance EMD

For measure the transformation quality of the aforementioned methods, a distance-based metric is used to calculate the similarity of the distribution between the observable variables and the construct. EMD is one of the methods that calculates the distance between classes of different levels by solving the domain alignment problem. EMD is a metric between two distributions and is defined as the minimum cost to transform one distribution into another, where there is a GD (ground distance) between the classes of the histogram [13]. The GD(i,j) is the distance that is calculated from the weight or maximum value between two classes of the compared histograms.

The EMD distance between the histograms **H** and **H′** is defined by Eq. 2.

$$EMD(H, H') = \frac{\sum_{i,j}^{N} d_{i,j} g_{i,j}}{\sum_{i,j}^{N} g_{i,j}} \tag{2}$$

At, $d_{i,j}$ indicates the dissimilarity between class i and class j of the compared histograms. For, $g_{i,j}$ is the optimal flow between the two distributions such that total cost $\sum_{i,j}^{N} g_{i,j}$ is minimized. The minimization process is subject to the following restrictions:

$$\sum_{i}^{N} g_{i,j} \leq H_j^{'} \quad (3)$$

$$\sum_{i}^{N} g_{i,j} \leq H_i^{'} \quad (4)$$

$$\sum_{i}^{N} g_{i,j} = \min\left(H_i, H_j^{'}\right) \quad (5)$$

4 Results

The data used for the variable transformation process are extracted from the case study "Individual and environmental factors that influence the dietary behavior of adolescents in low and middle-income settings in urban and rural areas of southern Ecuador" (10). The data obtained are part of a non-experimental design study that was applied in a cross-sectional manner. The research approach is mixed due to the use of quantitative and qualitative variables. The population analyzed were children and adolescents from 10 to 16 years of age from the city of Cuenca and Nabón. The sample size is 751 subjects and 77 variables. The factors and/or patterns were measured by collecting sociodemographic attributes (age, sex, geographic location and socioeconomic status), anthropometric measurements, surveys on dietary behavior and individual and environmental factors that influence dietary practices.

4.1 Variables Assignment

For data analysis, RStudio statistical software and the "emdist" library were used. Information was extracted from the database pertaining to the concepts mentioned in Sect. 2. The result of this extraction was 751 observations of 29 variables. Each variable is a survey question to which a label is assigned for its manipulation. Below are the questions associated with their respective labels and their established numerical assignment.

- **Self-efficacy**
  Self_eff1: How difficult is it for you to eat healthy every day?
  Self_eff2: How difficult is it for you to eat healthy at home?
  Self_eff3: How difficult is it for you to eat healthy at school?
- **Perceived benefits**
  Ben_bi1: If I eat healthy does it help me control my body weight?
  Ben_bi2: If I eat healthy does it make me feel good?
- **Habit Strength**
  Hab_1r: Do I eat snacks or fast food when I watch TV?
  Hab_2r: Do I eat snacks or fast food when I go out with friends?
  Hab_3r: Do I eat snacks or fast food when I go to sports training?
  Hab_4r: Do I eat snacks or fast food when I go on a family trip?

- **Perceived Barriers**
  Bar_1: Is unhealthy food cheaper?
  Bar_2: healthy food is not available when I have to eat?
  Bar_3: healthy food doesn't taste good?
  Bar_4: Have I been eating fast food since I was a very young child?
  Bar_5: Do my parents don't have time to cook healthy food?
  Bar_6: Does body needs unhealthy food?
  Bar_7: breaks at school are too short to eat healthy?
  Bar_8: Do I eat unhealthy because I want to eat the same food as my friends?
- **Parental permission**
  Prul_1: My parents let me eat fast food and snacks whenever I want?
  Prul_2: Do my parents let me drink soft drinks whenever I want?
- **Accessibility to healthy foods**
  Acc_1: My family can't afford to buy healthy food?
  Acc_2: Is there no weekly market in my neighborhood?
  Acc_3: Does the healthy food in my house go to waste?
  Acc_4: Are the places that sell healthy food far from my home?
- **School Support**
  Sch_rul1r: How often does your school encourage you to eat healthy?
  Sch_rul2r: How often does your school provide you with information regarding healthy eating?
- **Components of dietary behavior**
  qfrruit_new: quantity of fruits
  qveggie_new: amount of vegetables
  qsoft_new: sugar-sweetened beverages
  qsnacks_new: consumption of unhealthy snacks.

4.2 Outcome Constructs

According to the case study, there are 11 resulting concepts. These 11 concepts are represented as vectors of size 1×677, that is, there are 11 vectors with 677 data each one of them. As part of the proposed methodology 3 types of transformation of variables are established, then as results of the transformation of variables 3 data vectors are obtained, each one with 11 variables of size 1×677. To visualize the results and the behavior of the data of the resulting concept vector, a histogram of frequencies of the Self-efficacy concept is synthesized in Fig. 3.

The histograms generated represent the behavior of the resulting concept data. On the vertical axis are numerical responses of the participants to the questions asked in the surveys or interviews. The vertical axis represents the number of data for each concept or vector. To validate this transformation, we used the EMD metric seen in Sect. 3. This metric helps us to verify the cost of transforming the question variables to the concept variable.

To validate the techniques used for the transformation of question variables to concept variables, the EMD metric was used as a value that indicates the cost of transforming one distribution into another. From the mathematical point of view there is no considerable variation in the transformation of distributions, the cost is low and ranges between 0.1588

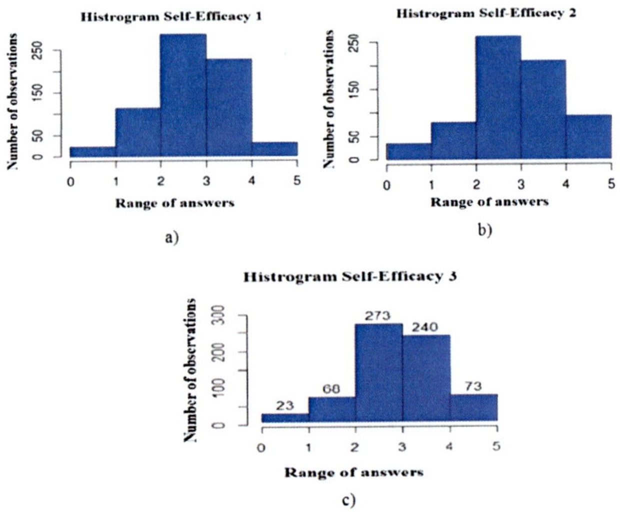

Fig. 3. a) Histogram of frequencies the concept self-efficacy with transformation technique 1, b) Histogram of frequencies the concept self-efficacy with transformation technique 2, c) Histogram of frequencies the concept self-efficacy with transformation technique 3.

and 0.5550 considering that the maximum threshold or maximum cost is 1 because the densities of the histograms are taken to enter the algorithm. Table 4 shows a summary of all the values obtained.

Table 4. Summary table with EMD values.

Transformation method 1				Transformation method 2				Transformation method 3			
Construct	Questions	EMD	Average	Construct	Questions	EMD	Average	Construct	Questions	EMD	Average
	Q1	0.208		C1	Q1	0.199		C1	Q1	0.114	
	Q2	0.402	0.3880		Q2	0.328	0.3860		Q2	0.204	0.3417
C1	Q3	0.554			Q3	0.631			Q3	0.708	
	Q1	0.180		C2	Q1	0.186	0.1647	C2	Q1	0.186	0.1647
C2	Q2	0.137	0.1588		Q2	0.143			Q2	0.143	
	Q1	0.557		C3	Q1	0.474		C3	Q1	0.428	
	Q2	0.386			Q2	0.456			Q2	0.570	
	Q3	0.461			Q3	0.532			Q3	0.666	
	Q4	0.383			Q4	0.453	0.5886		Q4	0.582	0.4857
	Q5	0.666			Q5	0.749			Q5	0.105	
	Q6	0.761			Q6	0.832			Q6	0.106	
	Q7	0.697			Q7	0.612			Q7	0.566	
C3	Q8	0.530	0.5550		Q8	0.601			Q8	0.863	
	Q1	0.167		C4	Q1	0.660		C4	Q1	0.411	
	Q2	0.253			Q2	0.808	0.7581		Q2	0.529	0.5011
	Q3	0.575			Q3	0.713			Q3	0.464	
C4	Q4	0.325	0.3298		Q4	0.851			Q4	0.601	
	Q1	0.360		C5	Q1	0.368	0.2984	C5	Q1	0.368	0.2984
C5	Q2	0.236	0.2984		Q2	0.229			Q2	0.229	
	Q1	0.539		C6	Q1	0.634		C6	Q1	0.778	
	Q2	0.586			Q2	0.610	0.5074		Q2	0.755	0.5532
	Q3	0.290			Q3	0.307			Q3	0.304	
C6	Q4	0.461	0.4690		Q4	0.479			Q4	0.375	
	Q1	0.380		C7	Q1	0.128	0.5123	C7	Q1	0.409	0.2644
C7	Q2	0.129	0.2541		Q2	0.897			Q2	0.120	
C8	Q1	0.000	0	C8	Q1	0.000	0	C8	Q1	0.000	0.000000
C9	Q1	0.000	0	C9	Q1	0.000	0	C9	Q1	0.000	0.000000
C10	Q1	0.000	0	C10	Q1	0.000	0	C10	Q1	0.000	0.000000
C11	Q1	0.000	0	C11	Q1	0.000	0	C11	Q1	0.000	0

5 Conclusions

The treatment of the data is fundamental to obtain good results, since the reliability of the data depends on it. There are different ways of associating or taking observable data to behavioral constructs; however, their reliability is not mathematically guaranteed. In this work, we proposed methods for the transformation of variables from categorical observable data to numerical data and with them to convert them into concepts under a mathematical support. When transforming a set of data, there is a risk that the data distribution of the transformation result will be totally altered and will not assimilate the expected behavior. By using quantitative transformation methods, it is possible to evaluate the similarity between the observable variables and the construct generated. In this case, the arithmetic mean and exponential function methods have a lower cost, i.e., there is greater similarity between the observable data and the construct.

References

1. Reed, Z.E., Micali, N., Bulik, C.M., Smith, G.D., Wade, K.H.: Assessing the causal role of adiposity on disordered eating in childhood, adolescence, and adulthood: a Mendelian randomization analysis. Am. J. Clin. Nutr. **106**(3), 764–772 (2017)
2. Olivari, C., Urra, E.: Sel-efficacy and health behaviors. Cienc. y Enefermeria (1), 9–15 (2007)
3. Barragán, W., Mahecha, L., Angulo, J., Olivera, M.: Percepción de consumidores frente a información nutricional de la carne bovina. Idesia, **36**(4), 35–43 (2018)
4. Mendoza-Parra, S., Merino, J.M., Barriga, O.A.: Identificación de factores de predicción del incumplimiento terapéutico en adultos mayores hipertensos de una comunidad del sur de Chile. Rev. Panam. Salud Publica/Pan Am. J. Public Heal. **25**(2), 105–112 (2009)
5. Fernández, V.: Relación entre la Capacidad de Absorción y las Estructuras Organizativas. Tesis Univ. Politécnica Catalunya, p. 287 (2004)
6. Benarroch, A., Pérez, S., Perales, J.: Factores que influyen en las conductas alimentarias de los adolescentes: Aplicación y validación de un instrumento diagnóstico. Electron. J. Res. Educ. Psychol. **9**(3), 1219–1244 (2011)
7. Morales, E.M., Maghioros, M.A., Obregón, A.M., Santos, J.L.: Adaptación y análisis factorial del cuestionario de conducta de alimentación del adulto (AEBQ) en idioma español. Arch. Latinoam. Nutr. **70**(1), 40–49 (2020)
8. Closas, A.H., Arriola, E.A., Kuc, C.I., Amarilla, M.R., Jovanovich, E.C.: Análisis multivariante, conceptos y aplicaciones en Psicología Educativa y Psicometría. Enfoques **XXV**(1), 65–92 (2013)
9. Huang, Y., et al.: Obesity and peripheral arterial disease: A Mendelian randomization analysis. Atherosclerosis **247**, 218–224 (2016)
10. Verstraeten, R., et al.: Individual and environmental factors influencing adolescents' dietary behavior in low- and middle-income settings. PLoS ONE **11**(7), e0157744 (2016)
11. Snavely, A.C., Harrington, D.P., Li, Y.: A latent variable transformation model approach for exploring dysphagia. Stat. Med. **33**(25), 4337–4352 (2014)
12. Li, S., Sari, Y.A., Kumral, M.: New approaches to cognitive work analysis through latent variable modeling in mining operations. Int. J. Min. Sci. Technol. **29**(4), 549–556 (2019)
13. Rubner, Y., Tomasi, C., Guibas, L.J.: The earth mover's distance as a metric for image retrieval. Int. J. Comput. Vis. **40**, 99–121 (1998). https://doi.org/10.1023/A:1026543900054

Check for updates

Smart City Planning with Geomatic Modeling of Lahar Evacuation Routes in the Northern Populated Area of Cotopaxi Volcano, Ecuador

Oswaldo Padilla Almeida[1], Theofilos Toulkeridis[1,2](✉), and Joaquín Bosque Sendra[3]

[1] Universidad de las Fuerzas Armadas ESPE, Sangolquí, Ecuador
ttoulkeridis@espe.edu.ec
[2] Universidad de Especialidades Turísticas UDET, Quito, Ecuador
[3] Departamento de Geología, Universidad de Alcalá, Geografía y Medio Ambiente. Antiguo Colegiode Málaga, Alcalá de Henares, Madrid, Spain

Abstract. The Cotopaxi Volcano is one of the most dangerous volcanoes worldwide due to its potential of the generation of voluminous lahars of dozens of millions of cubic meters capable of destroying infrastructure and endangering a lot of people living near major river drainages of this volcano. Our study describes such circumstances in the northern side of Cotopaxi Volcano and how we pro- pose to reduce the vulnerability of the public with new evacuation methods. Therefore, we have used geomatic tools, in order to shorten evacuation ways and directions. Based on the results, we determined different spatial variables or geographic coverage of the described and highlighted the main points of interest in each of them. The location of initial evacuation points of the population was determined within the lahar travel area being along the road axes. With these points we calculated security checkpoints outside the area lahar with additional margin. For this process the impedance was determined according to the average speed of a person in case of evacuation. In areas where the evacuation time has been longer than the arrival time of the lahar, vertical rather than horizontal evacuation points were determined by evaluating its coverage area depending on the time needed for the population to be safe.

Keywords: Lahar · Evacuation routes · GIS · Algorithm of Dijkstra · Ecuador

1 Introduction

Volcanoes and associated hazards have been responsible for the death of hundreds of thousands of people in the last two centuries worldwide [1, 2], destroying a variety of strategic infrastructure throughout the world, changing the local and global climate [3–7] as well as nearby landscapes [8–11]. One of the most lethal volcanic hazards are lahars, as they are more devastating in terms of cumulative fatalities than all other volcanic hazards including pyroclastic flows [12, 13].

The Cotopaxi Volcano located in the western cordillera of the Ecuadorian Andes is known to have had a vast history of big eruptions, far-reaching ash emissions and

S. Berrezueta and K. Abad (Eds.): *Doctoral Symposium on Information and Communication Technologies - DSICT*, LNEE 846, pp. 74–88, 2022.
https://doi.org/10.1007/978-3-030-93718-8_7

most importantly lahar generations, which may surpass 150 million m^3 [14–22]. Lahar generation may be caused by either by the collapse of an eruptive column or by the effect known as boiling over, both produce pyroclastic flows, which are able to melt part of the glacier´s surface [15, 18, 19]. As the volcano awakens, it is fundamental to know the flow directions of the generated lahars in order to reduce risks, vulnerabilities and losses similar and/or worse than the ones that destroyed Armero town, Colombia in 1985 killing some 23,000 citizens [12, 23–25]. The decision of how and when to evacuate potential hazard zones threatened by lahars may be fundamental for the survival of villages and people living near the volcano [26–31]. As the probability of a reactivation of the volcano is relatively high with some 72% in 2016 [31], the question is not if but when the volcano will generate big explosions of VEI $>$ 3 (up to VEI 5) and subsequent generations of deadly, far-reaching lahars.

Therefore, the main aim of this smart city study has been a novel approach to evaluate potential evacuation times and evacuation behavior in order to reduce life losses in a densely populated area of the Valley de los Chillos in the preferred lahar hazard zones in the northern flow direction of Cotopaxi volcano. The geographic tools shall allow to reduce time spans between evacuation points towards safe zones outside the lahar hazard areas and where applicable towards higher elevations within the hazard zones.

2 Geological Setting and Volcanic History of Cotopaxi Volcano

The northern cordilleran Andes in Ecuador are part of the 7000 km long classic example of an active continental margin along the South American continent, with several volcanic sequences of Mesozoic and Cenozoic ages. More than 250 volcanoes are exposed in the Ecuadorian part of the Northern Andean Volcanic Zone (NAVZ) of which the 5897 m a.s.l. high Cotopaxi, is one of the twenty considered active volcanoes in the country as a result of the subduction of the oceanic Nazca plate below the South American continent [32–35]. In the area of Ecuador, the Nazca Pacific plate is subducted at an angle slightly oblique to the southern American continent, producing an overall active tectonic regime with transpression due to its convergence. The consequences of this subduction are four morphologically distinctive volcanic chains [35].

The Cotopaxi stratovolcano in Ecuador, which has been formed in the Pleistocene, is located some 60 km south of Quito and represents a natural laboratory for the assessment of volcanic hazards (Fig. 1). The current volcano consists of two craters, one older and a more recent one. The more recently created crater is snow-covered and is the site of volcanic vent activity. Two partial sector collapses occurred some 4,600 and 2,300 years ago forming debris avalanches and leaving some hummocks behind, while the usual volcanic activity is expressed by precipitation of ash and pumice as well as by pyroclastic flows, high gas emissions, lava flows and volcanic mud flows, known as lahars [36]. The magmatic activity of the volcano in historical times is well documented [37–45] and the lahars produced by the volcano had considerably affected villages, cities and infrastructure in its surrounding. Some 19 eruptive phases have been registered and dated giving a re-occurrence of its activity every 117 $\pm$ 70 years over the last 2200 years [14, 16]. The last four volcanic phases with the generation of lahars and subse- quent destruction of nearby villages occurred in 1534, 1742, 1768 and [14, 16].

The best-documented event took place in 1877 and ended up killing approximately 1,000 people in the area [2, 40]. Glacier melting facilitated the creation of a lahar, which roared down the mountain at speeds of up to 70 km/h [15]. Although the glacier is currently retreating, circumstances have changed since the 1877 lahar. Thus, although a future lahar would be smaller than that of 1877, the fact that many more people are living within the area suggests that the loss of life could be much higher than in the past. Whereas about 30,000 people were living in the danger zone in the late 1800s, currently the number is greater than 500,000 [31]. The most dangerous areas would be the Southern, Eastern, and Northern sides due to flow patterns (Fig. 1). Unfortunately the nearest cities would be hit within 30 minutes from the formation of a lahar, with little warning for evacuation [15, 16]. In addition to these highly destructive volcanic events, similar to 1877, some additional 59 other minor volcanic events have been recorded between 1532 and 2015, 27 of which included lava flows and/or minor lahars besides regular gas emission and precipitation of pyroclastic material [15, 16, 22, 46–48]. Based on statistical data from the last 19 eruptive phases over 2200 years, the probability of a new occurrence in the year 2016 has been of about 72%, while the re-occurrence of a minor volcanic event, like the 59 sporadic explosions has been much higher [16, 31, 49, 66].

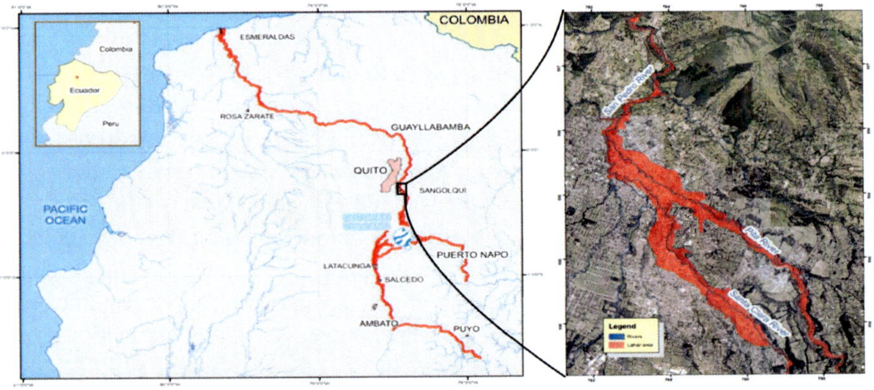

Fig. 1. Inlet upper left: Ecuador s position in northwestern South America. Left: Ecuador with the trajectory of lahars of the volcano Cotopaxi. Right: the study area of the Chillos Valley with the lahars of the Rivers Pita and Santa Clara.

In the northern drainage zone of the Cotopaxi volcano, cities like Sangolquí and San Rafael will be hit in less than half an hour after the generation of the lahars [15]. The rivers in such zone, which will be flooded by the lahars are Salto, Pita and Santa Clara (Fig. 1). The evaluation of the economic loss of the lahars yielded a total loss of approximately 17 billion USD representing around 16.8% of the Ecuadorian PIB of which 81% are restricted to the northern side of the [31, 51, 52].

2.1 Description of the Hazard Zone and Associated Traffic

For the modeling and calculation of ideal evacuation routes it is fundamental to include and evaluate the vehicular behavior, traffic and people density, road infrastructure and accessibility in the hazard zone [53, 54]. The most regular evacuation behavior is horizontal in order to escape the incoming volcanic lahar towards the closest point outside the known flowing path. Nonetheless, we will also consider vertical evacuation routes where the escape time may take longer time than the impact of a lahar, including hereby lahar-impact resistant, tall buildings where a less-lasting vertical evacuation will be helpful. Like in many cities nationwide, a significant population growth during the last decades increased the population density also in the cities of San Rafael and Sangolquí, without a contemporary corresponding improvement of the road infrastructure, which lacks in quality (both in design and maintenance) causing severe congestion problems in clearly identifiable sites. Therefore, the problems of traffic congestion have been added as an additional risk factor in case of mobilization of the population being an obstacle for potential evacuation routes for the public.

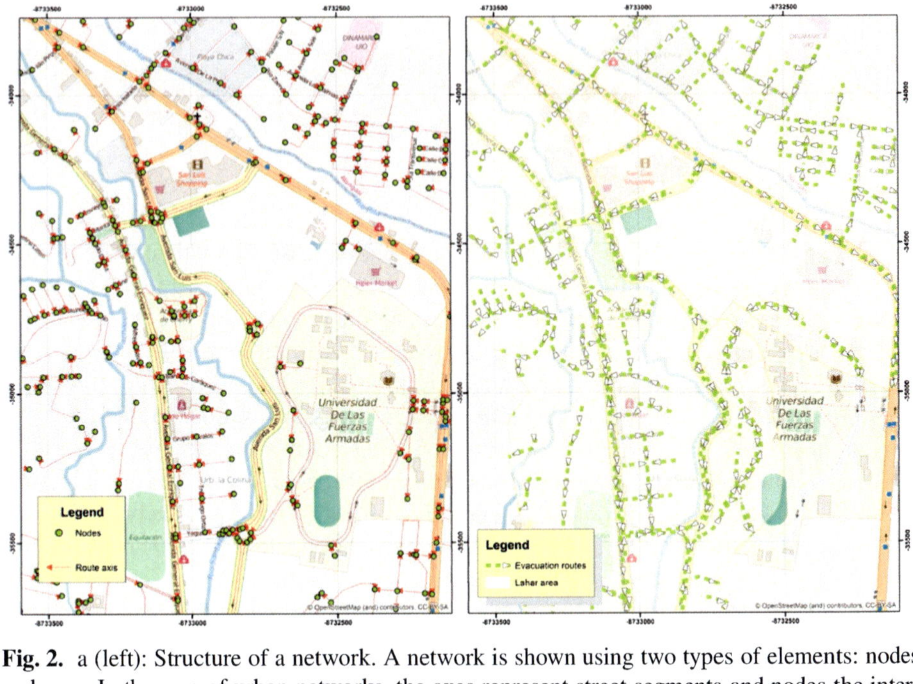

Fig. 2. a (left): Structure of a network. A network is shown using two types of elements: nodes and axes. In the case of urban networks, the axes represent street segments and nodes the intersection between two or more axes. b (right): Evacuation routes in case of a lahar through a network in the lahar area of Sangolquí-San Rafael.

Currently, the critical points where the most traffic chaos in the Valley of the Chillos are places known as El Triangle, the Tourism Park, El Choclo and in the area of the San Luis Mall, which add difficulties in mobility leads to move to Quito [55]. The most important roads are the route E35 (General Pintag Avenue), the General Rumiñahui Avenue both directed north-south and the Intervalley way perpendicular to them demonstrating the few alternatives of the citizens to reach safe zones in case of an emergency generated by lahar flows (Fig. 2). All these roads are highly trafficked almost at all times of the day but severe congestion is generated especially between 6:30 a.m. to 9:00 a.m., from 12:00 pm to 1:30 pm, and from 3:30 pm to 7:00 pm. [55]. Even more congestion occurs during holidays and local celebrations and less during school vacation times. However, the situation gets worse considering the presence of poor road quality, narrow streets often even blocked with illegally parked vehicles, two-sided traffic on the same road, lack of sidewalks or frequently obstructed by cables, posts, ads, informal commercial activity, and speed bumps, making it very difficult or even impossible to use strollers or wheelchairs. Furthermore, there is inefficient and poor public transport service. Out of the mentioned, the conflicting sites with frequent traffic jams are the Tourism Park, the bridge over the Santa Clara River ("San Luis Mall" sector), the surrounding of the ESPE university, the Triangle sector, the San Luis Shop- ping sector, the bridge in front of the Farina College and the roundabout of the River Mall (Fig. 2). Based on the last census of 2010, some 82,000 citizens live within the lahar zones in the study area, while others several thousand citizens are temporarily present during their daily socio-economic activities.

2.2 Methodology

a) *Geographic Information System*

In the spatial environment, the road network plays a decisive role. Even more so when natural hazards are present, counting as a factor that increases the vulnerability of the population and infrastructure. As the entire road infrastructure facilitates and provides access to different sites, an evacuation model using the road system is vital to manage any type of a potentially catastrophic event related to natural hazard occurrences. In this respect, Geographic Information Systems (GIS) are useful tools that allow the combination of a variety of variables creating different scenarios in order to find the most suitable one solving an existing reality situation in the study area, towards the possible occurrence of a potential future catastrophic phenomenon. Given the available knowledge of the territory based on the disintegration of the different layers acting on it, we have chosen the GIS as a tool for designing a model of horizontal and vertical evacuation as well as for risk management and security for a given event.

Therefore, for the best possible results in the calculation of accessibility and modeling of eventual evacuations of an area such as the one in San Rafael and Sangolquí threatened by lahars of the Cotopaxi volcano, it requires some data entry and information of high quantities and quality, for which GIS allows such needed technological conditions. These decision-making tools, among other benefits allow the graphic and visual understanding of the studied hazards through the generation of a variety of maps. Among the different application areas of GIS is network analysis. That is a fundamental tool for design, planning, and community development, as it defines a network as an interconnected system of linear elements and connection points that form a structure which may move people, energy, information and property [56]. A network is given when two types of elements are used, namely nodes and axes. In the case of urban networks, the axes represent street segments while nodes represent the intersection be- tween two or more axes [57].

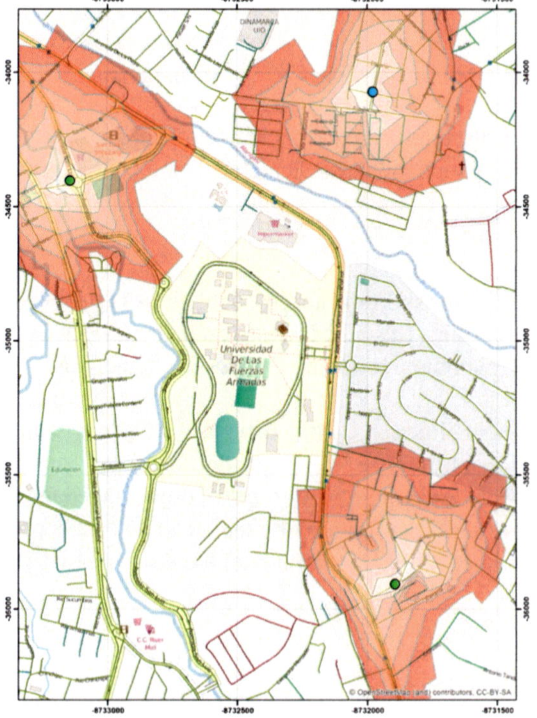

Fig. 3. Calculation of service areas. From the lightest to the darkest red are pointed out the circulation times ranging from one to six minutes towards the outermost ring in which they are represented.

A network analysis can be useful for a number of urban applications such as transportation route planning [58], flow of pedestrian traffic in the city to determine the busiest routes [59, 60]. Location and distribution of advertising, commercial and service establishments in urban environments [61] or in risk assessment [62]. The main objective of network analysis is to optimize the use of resources [63] through the determination of the shortest routes or the optimal routes based on the time of evacuation, which additionally will determine the number of people who may be saved before the possible impact of the natural phenomenon that threatens the population. However, a road net- work should have some basic attributes such as street name, maximum allowable flow rate, length, order, among others, and mainly impedance, which will depend on the remaining attributes. The impedance (resistance or friction to movement) as the resistance offered by each road axis to free movement measured in units of time [64, 65].

The impedance becomes a chastisement factor towards mobilization time for the route segment, so that two segments with an equal distance may have two different impedance values, resulting in obtaining two times of the different routes. The first factor affecting the impedance is the distance. Further factors are such as the slope [66], the number of individuals or vehicles on the road, the number of lanes, the road conditions as well as the type of road. Each of these elements will give a different friction factor value, which will comprise all individual friction, providing the final value of mobilization expressed in the Eq. 1.

$$Mobilization = \frac{Length}{velocity} \times impedance \quad (1)$$

Where the length is the distance in meters of each road segment and the speed is expressed in meters per minute at which the network elements are on the move (Fig. 2 and 3).

b) *Algorithm of Dijkstra.*

The most used algorithm for short distances calculations is the Dijkstra algorithm [67–70]. This algorithm determines also the lowest cost path between two points, calculating the costs accrued between the intermediate nodes of the network, making iterations to find the shortest path [71, 72] Dijkstra's algorithm solves the problem of the shortest path in a network without directions, with no negatives and not being weighted. In order to use it in our study and purpose, this algorithm has been modified to respect user settings such as one-way restrictions, u-turn restrictions, impedances of nodes, barriers and restrictions across the street minimizing the attribute cost specified by the user. In addition, the algorithm should be able to model locations anywhere along an axis, not only in or along joints.

3 Results and Discussion

In order to formulate the evacuation scenarios, it is important to define the first two elements for the calculation of evacuation times being the security and initial evacuation points. Defined minimum security points correspond to points of intersection of the main roads to the closest limit of the evacuation area, being the places where the population will be without any imminent risk of being affected by the incoming lahar. The starting points are placed at each corner points searching for more points for an appropriate access on every square, corresponding to the or initial evacuation points (Fig. 4). With the given elements such as security points, evacuation points and main roads, the different evacuation scenarios were modeled. The first variable used for calculating evacuation times corresponds to the time of arrival of the lahar [15, 16].

The two different evacuation scenarios, which have been considered for the study area are those declared by the monitoring staff due to a sudden change in the volcanic behavior and the subsequent declaration of an orange alert by the authorities, who will order a mandatory evacuation. In such situation, there will be enough time to evacuate the hazard areas without any difficulties [28, 73–75, 75, 77]. The second scenario is based on a sudden generation of lahars, which will impact the hazard area in a relatively short time. In such a case, due to a potential early alert system, there will be a very reduced time span to warn and evacuate the population in danger. Therefore, proper and early planning is essential making this approach of modeling the shortest evacuation routes necessary based on the established scenarios.

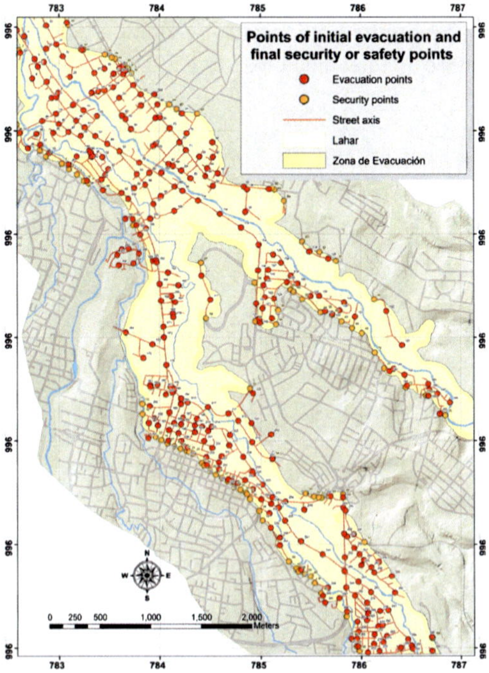

Fig. 4. Calculated points of initial evacuation and final security or safety points

Assuming the worst situation, evacuation scenarios for the given hazard zone are limited to be by vehicle (car or similar), by foot or a combined evacuation (foot-vehicles). The evacuation type using a vehicle is not the most recommended one, if evacuation time is critical and the transported mass is a high amount of people. That is not the case for the arrival of a tornado or hurricane, which allows a time of one to two days for the evacuation and the use of vehicles. In the study area with the use of vehicles the evacuation may take as long as 54 min for certain zones, way too long to escape an incoming lahar (Fig. 5a). Such times have been measured and calculated taking as a parameter for calculating the flow velocity in low extreme conditions at 5 km/h, and even less at times when traffic jams occur and circulation is even more complicated.

On the other side, the combined evacuation by vehicle and by foot may also be problematic, as it most probably provokes traffic jams due to the large number of vehicles to be used and the high amount of people in the evacuation routes. Such a situation generates car users to leave the vehicles and continue by foot enlarging the problem of traffic jams, a situation which occurred during an earthquake and a supposed tsunami in March 2014 in Arica and Antofagasta, northern Chile [78]. In San Rafael and Sangolquí, this may take for certain spots longer than 40 min, although shorter than the previous scenario, still too long to survive for some groups in such an area. In particular, there are two major spots where jams occur where vehicular traffic is void. The first one is the Mall San Luis towards the Triangle and the second is off Sangolquí to the park of San Rafael (Fig. 5b). Both areas become areas of gridlock as they do not have enough alternate routes to vent, becoming bottlenecks, a situation that is evident daily in peak traffic hours and even more at dates like close to Christmas.

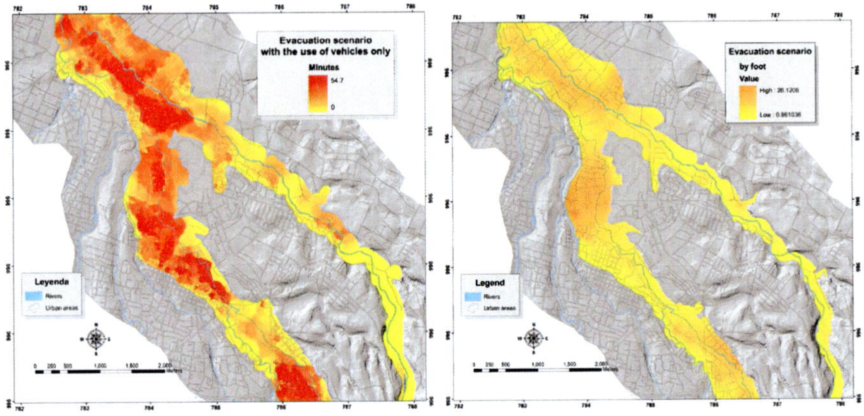

Fig. 5. a (left) Evacuation scenario with the use of vehicles only. b (right). Evacuation scenario with the use of no vehicles at all, meaning by foot.

Therefore, the most suitable type of evacuation in a sudden type event will be the evacuation by foot only, allowing a more fluid way of leaving hazard areas without any restrictions like those generated by vehicular movements. This evacuation needs mostly with few exceptions (Table 1) a time less than the arrival of a lahar if almost perfect conditions are respected (Fig. 5b).

However, if these conditions are not given as desired, then the evacuation time by foot is on the limit of the impact time of the incoming lahar. Therefore, in this respect we have considered two ways to evacuate by foot, as there were exclusively horizontally or horizontally combined with a vertical evacuation where possible (Fig. 6a, b). The horizontal evacuation is known to be usually that kind where people use the known escape routes towards safe areas within the shortest distances and times. Nonetheless, as often the horizontal distances towards a safe zone may be too far and too long in respect to the incoming hazard like the lahar in our study. Therefore, the combination of a horizontal movement on the way to an elevated safe area like certain buildings that meet particular conditions and are capable of withstanding the event may be an adequate solution for the people in danger.

In order to fulfill conditions of safe zones in vertically accessible zones, we evaluated the higher buildings in these areas, meaning that the building height is higher than the maximum height of the lahar. Furthermore, such high structures need to have their lower access always open and accessible and be designed in a way in which their ascent routes are spacious and have the respective securities. Additionally, such buildings also must provide large enough space to accommodate temporarily the evacuated persons. In this respect, we identified 55 buildings in the hazard zone. Such zones of vertical safety reduce the evacuation time drastically, as persons even in not adequate physical conditions, who would need 15-20 minutes to reach a safe zone outside the drainage filled with the lahar flow, may need up to 3´40´´at most to be safe within the hazard zone.

Table 1. Number of persons of the areas that are unable to perform a horizontal evacuation in time prior arrival of the lahar impact. Total habitants: 1997 people

		Population density	Area	Persons
ID	Area (m^2)	(person/Km^2)	(Km^2)	(Area x Pop. Dens.)
1	67.17		0.0001	0
2	593812.4		0.5938	841
3	143.85	1416.06	0.0001	0
4	669418.3		0.6694	948
5	5666.99		0.0057	8

Of these 55 buildings, we have chosen three, one for each vertical evacuation zone for further analysis: the San Luis Shopping Center, the Sun Valley Building, and the Balcony Valley Building (Fig.7). These buildings, called vertical safety points, have wide access, which will allow the rise of the evacuated persons safely and relatively quickly (Fig. 7). The streets and avenues that reach each security checkpoints are in relatively good condition, which would provide adequate accessibility to the areas near these points. While the Balcony Valley Building, which is located nearby the War Academy and the San Jorge Military Fort, is able to serve as a shelter for the 885 persons who live there, the San Luis Shopping Center has even more space. Being one of the biggest shopping malls of the country and situated on the spot where two lahars of the rivers Santa Clara and Pita meet, it is able to serve for a shelter of up to 2500 persons. Finally, the Sun Valley Building at the Triangle zone, comprises some 526 people living within this evacuation coverage. Here are sites like the Valley Mall, the Valley Plaza Mall and also an area with a high density of further businesses and shops. It is the busiest area of San Rafael, at the confluence of the main routes arriving and departing the Valley of the Chillos, like the General Rumiñahui Highway and the Via Tingo- Conocoto.

These three particularly described vertical safety points are representative to the other but smaller 52 residual high buildings, within the hazard zone. Each one of the 55 high structures represent the area of influence of the buildings or vertical safety points, based on mobilization times expressed by a polygon on the ground. Such coverage areas have been obtained based on the times of each of the scenarios previously determined. These areas are calculated based on the concept of "service areas" representing the area covered by a temporary measure equidistance from a given point. As a final result con sidering the shortest horizontal and vertical evacuation routes, we configured the most ideal evacuation directions for the hazard zones as illustrated in Fig. 2b.

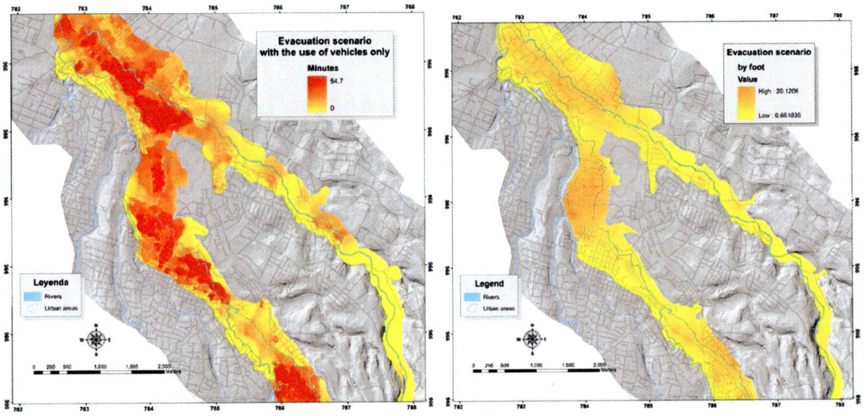

Fig. 6. a (left)**:** Evacuation scenario with the mixed use of vehicles and by foot. **b** (right): Same scenario where data are shown in respect to evacuation times of the public being in risk. Two distinctive zones are here identified, in which evacuation times are higher than those of the arrival of a lahar, being precisely those areas where a vertical evacuation is recommended.

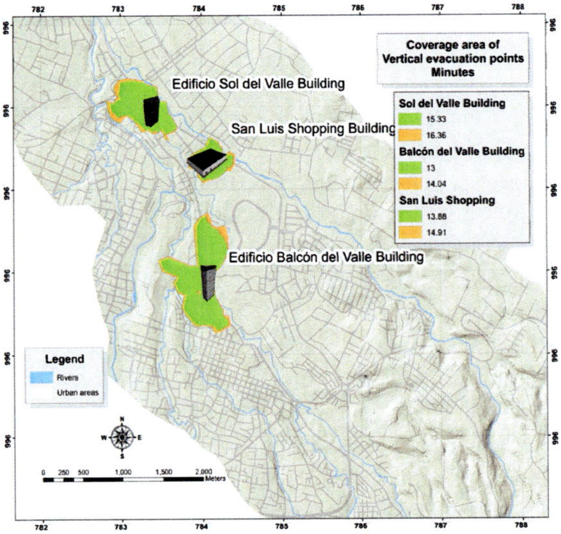

Fig. 7. Evacuation coverage areas for each security checkpoint of the three buildings mentioned in the text.

4 Conclusions

Risk assessment has been applied in a hazard zone threatened by the impact of lahars of the volcano Cotopaxi with the use of geomatic tools such as geographic information techniques and algorithms to determine the best and shortest evacuation distances.

These prevention techniques with a detailed network analysis enabled us to determine horizontal and eventual vertical evacuation sites, where possible to reduce even more evacuation times prior to lahar impact.

The study with the obtained results demonstrates a powerful tool for the corresponding authorities such as government entities responsible for risk management and the public alike as an input for proper land use planning and enabling to save lives. The data of our study allows us to channel evacuation ways and avoid uncontrolled escapes during a potential volcanic crisis.

References

1. Peterson, D.W.: Volcanic hazards and public response. J. Geophys. Res. **93B5**, 4161–4170 (1988)
2. Tanguy, J.-C., Ribière, C., Scarth, A., Tjetjep, W.S.: Victims from volcanic eruptions: a revised database. Bull. Volcanol. **60**(2), 137–144 (1998). https://doi.org/10.1007/s004450050222
3. Hofmann, D.J., Rosen, J.M.: Sulfuric acid droplet formation and growth in the stratosphere after the 1982 eruption of El Chichon. Science **222**, 325–327 (1983)
4. Self,S., Rampino, M.R.: The relations between volcanic eruptions and climate change: still a conundrum. Eos **69,** 74–75, 85–86 (1988)
5. Pinatubo Volcano Observatory Team: Lessons from a major eruption: Mt. Pinatubo, Philippines. EOS **72** (49), 545–555 (1991)
6. Hansen, J., Lacis, A., Ruedy, R., Sato, M.: Potential climate impact of Mount Pinatubo Eruption. Geoph. Res. Lett. **19**, 215–218 (1992)
7. Briffa, K.R., Jones, P.D., Schweingruber, F.H.: Influence of volcanic eruptions on Northern Hemisphere summer temperature over the past 600 years. Nature **393**, 450–455 (1998)
8. Blong, R.J.: Volcanic Hazards, p. 424. Academic Press, Sydney, Australia (1984)
9. Jago, L.C., Boyd, W.E.: How a wet tropical rainforest copes with repeated volcanic destruction. Quatern. Res. **64**(3), 399–406 (2005)
10. Scott, W.E., Nye, C.J., Waythomas, C.: F: August 2008 eruption of Kasatochi volcano, Aleutian Islands, Alaska—resetting an island landscape. Arctic Antarctic Alpine Res. **42**(3), 250–259 (2010)
11. Aviles-Campoverde, D., Chunga, K., Ortiz-Hernández, E.: Seismically induced soil liquefaction and geological conditions in the city of Jama due to the Mw7.8 Pedernales earthquake in 2016, NW Ecuador. Geosciences **11**, 20 (2021)
12. Sigurdsson, H., Carey, S.: Volcanic disasters in Latin America and the 13th November 1985 eruption of Nevado del Ruiz volcano in Colombia. Disasters **10**(3), 205–216 (1986)
13. Rodolfo, K.S.: The Hazard from Lahars and Jökulhlaups. In: Sigurdsson, H., et al. (eds) Encyclopedia of Volcanoes, pp. 973–995. Academic Press, Narragansett (1999)
14. Barberi, F., Coltelli, M., Frullani, A.: Chronology and dispersal characteristics of recently (last 5000 years) erupted tephra of Cotopaxi (Ecuador): implications for long-term eruptive forecasting. J. Volcanol. Geoth. Res. **69**(3), 217–239 (1995)
15. Aguilera, E., Pareschi, M.T., Rosi, M.: Risk from lahars in the northern valleys of Cotopaxi Volcano (Ecuador). Nat. Hazards **33**(2), 161–189 (2004)
16. Aguilera, E., Toulkeridis, T.: El Volcán Cotopaxi – Una amenaza que acecha. COSUDE, CGVG-USFQ, ESPE, 52 pp. FOES, Quito (2005)
17. Pistolesi, M.: Historical analysis and stratigraphy of the post-XII century pyroclastic activity at Cotopaxi volcano, Ecuador Implication for lahar hazard assessment. Plinius **34**, 131–138 (2008)
18. Pistolesi, M., Cioni, R., Rosi, M.: Evidence for lahar-triggering mechanisms in complex stratigraphic sequences: the post-twelfth century eruptive activity of Cotopaxi Volcano. Ecuador. Bull. Volcanol. **75**(3), 1–18 (2013)

19. Pistolesi, M., Cioni, R., Rosi, M.: Lahar hazard assessment in the southern drainage system of Cotopaxi volcano, Ecuador: Results from multiscale lahar simulations. Geomorphology **207**, 51–63 (2014)
20. Toulkeridis, T., Arroyo, C.R., Cruz D'Howitt, M.: Evaluation of the initial stage of the reactivated Cotopaxi volcano - analysis of the first ejected fine-grained material. Nat. Hazards Earth Syst. Sci. **3**(11), 6947–6976 (2015)
21. Vaca, A.V., Arroyo, C.R., Debut, A.: Characterization of fine-grained material ejected by the Cotopaxi volcano employing X-ray diffraction and electron diffraction scattering techniques. Biol. Med. (Aligarh) **8,**280 (2016)
22. Toulkeridis, T., Zach, I.: Wind directions of volcanic ash-charged clouds in Ecuador 2017: implications for the public and flight safety. Geomat. Nat. Hazards Risks **8**(2), 242–256 (2017)
23. Naranjo, J.L., Sigurdsson, H., Carey, S.N., Fritz, W.: Eruption of the Nevado del Ruiz volcano, Colombia, on 13 November 1985: tephra fall and lahars. Science **233**(4767), 961–963 (1986)
24. Thouret, J.C.: Effects of the November 13, 1985 eruption on the snow pack and ice cap of Nevado del Ruiz volcano, Colombia. J. Volcanol. Geothermal Res. **41**(1), 177–201 (1990)
25. Pierson, T.C., Janda, R.J., Thouret, J.-C.: Perturbation and melting of snow and ice by the 13 November 1985 eruption of Nevado del Ruiz, Colombia, and consequent mobilization, flow and deposition of lahars. J. Volc. Geotherm. Res. **41**, 17–66 (1990)
26. Paton, D., Flin, R.: Disaster stress: an emergency management perspective. Disast. Prev. Manag. Int. J. **8**(4), 261–267 (1999)
27. Schmincke H.U.: Volcanism. 326 p. Springer, Heidelberg (2004). https://doi.org/10.1007/1-4020-4496-8_337
28. Haynes, K., Barclay, J., Pidgeon, N.: The issue of trust and its influence on risk communication during a volcanic crisis. Bull. Volcanol. **70**(5), 605–621 (2008)
29. De la Cruz-Reyna, S., Tilling, R.I.: Scientific and public responses to the ongoing volcanic crisis at Popocatépetl Volcano, Mexico: importance of an effective hazards-warning system. J. Volcanol. Geoth. Res. **170**(1), 121–134 (2008)
30. Barclay, J., Haynes, K., Mitchell, T.: Framing volcanic risk communication within disaster risk reduction: finding ways for the social and physical sciences to work together. Geol. Soc. Lond. Special Public. **305**(1), 163–177 (2008)
31. Rodriguez, F., Toulkeridis, T., Padilla, O.: Economic risk assessment of Cotopaxi volcano Ecuador in case of a future lahar emplacement. Nat. Hazards **85**, 605–618 (2016)
32. Lonsdale, P.: Ecuadorian subduction system. Bull. Am. Assoc. Pet. Geol. **62**, 2454–2477 (1978)
33. Barberi, F., Coltelli, M., Ferrara, G.: Plio-Quaternary volcanism in Ecuador. Geol. Mag. **125**, 1–14 (1988)
34. Freymuller, J.T., Kellogg, J.N., Vega, V.: Plate motions in the north Andean region. J. Geophys. Res. **98**, 21853–21863 (1993)
35. Toulkeridis, T.: Volcanes Activos Ecuador, p. 152. Santa Rita, Quito, Ecuador (2013)
36. Mothes, P., Hall, M., Janda, R.: The Enormous Chillos Valley Lahar: an ash-flow gen- erated debris flow from Cotopaxi volcano. Ecuador. Bull. Volcanol. **59**, 233–244 (1998)
37. La Condamine Ch. M.: Diario del Viaje al Ecuador, republ. In 1986, Quito, Publitécnica: 221 pp. (1751)
38. von Humboldt, A.: 1837–1838: Geognostische und physikalische Beobachtungen über die Vulkane des Hochlandes von Poggendorff s. « Ann. Phys. Chem », Bd. **40**, 161–193 and Bd. **44**, 193–219
39. Reiss, W.: Über Lavaströme der Tungurahua und Cotopaxi. Zeitschr. Dt. Geol. Ges. **26**, 907–927 (1874)
40. Sodiro L.: Relación Sobre la Erupción del Cotopaxi Acaecida el Día 26 de Junio, de 1877, Quito, Imprenta Nacional, Ecuador: 40 pp (1877)

41. Stübel, A.: Die Vulkanberge Ecuadors, p. 556. A, Berlin (1897)
42. Whymper E.: Travels Amongst the Great Andes of the Equator, 456 pp. Pere grine Smith Books, Salt Lake City (1982)
43. Wolf, T.: Memoria Sobre el Cotopaxi y su Ultima Erupción Acaecida el 26 de Junio de 1877, 48 pp. Guayaquil, Imprenta de El Comercio, Ecuador (1878)
44. Wolf T.: Crónica de los Fenómenos Volcánicos y Terremotos en el Ecuador, Quito, Imprenta de la Universidad Central, 167 pp (1904)
45. Reiss W., Stubel A.: Das Hochgebirge der Republik Ecuador ii: Petrographische Untersuchungen des Ostkordillere, Berlin (1869–1902)
46. Barberi, F., Caruso, P., Macedonio, G.: Reconstruction and numerical simulation of the 1877 eruption of Cotopaxi volcano (Ecuador). Acta Vulcanológica **2**, 35–44 (1992)
47. Aguilera, E., Calatrava, L., Pareschi, M.T.: Numerical simulation of the lahars triggered by the 1877 eruption of Cotopaxi Volcano (Ecuador). Trans. Ecol. Environ. **12**, 205–214 (1986)
48. Toulkeridis, T., Seqqat, R., Torres, M.: COVID-19 Pandemic in Ecuador: a health disparities perspective. Revista de Salud Publica de Colombia **22**(3), 1–5 (2020)
49. Toulkeridis, T., Jacome, A., Mato, F.: Ethics, policy, and risk assessment of the Cotopaxi volcanic crisis in Ecuador—vulnerable society versus unprepared volcanic monitoring staff and authorities. In: Acevedo, R.D., Frías, J.M. (eds.) Geoethics in Latin America. TLASBS, pp. 153–170. Springer, Cham (2018). https://doi.org/10.1007/978-3-319-75373-7_11
50. Herrera-Enríquez, G., Toulkeridis, T., Castillo-Montesdeoca, E., Rodríguez-Rodríguez, G.: Critical factors of business adaptability during resilience in Baños de Agua Santa, Ecuador, due to volcanic hazards. In: Botto-Tobar, M., Cruz, H., Díaz Cadena, A. (eds.) CIT 2020. AISC, vol. 1327, pp. 283–297. Springer, Cham (2021). https://doi.org/10.1007/978-3-030-68083-1_22
51. Echegaray-Aveiga, R.C., Rodríguez, F., Toulkeridis, T.: Effects of potential lahars of the Cotopaxi volcano on housing market prices. J. Appl. Volcanol. **9**, 1–11 (2019)
52. Robayo N.A., Llorca, J., Toulkeridis, T.: Territorial, population and economic analysis of a potential volcanic disaster in the City of Latacunga, Central Ecuador Based on GIS Techniques – Implications and Potential Solutions. In: Rodriguez Morales, G., Fonseca, C.E.R., Salgado, J.P., Pérez-Gosende, P., Orellana Cordero, M., Berrezueta, S. (eds.) Information and Communication Technologies. TICEC 2020. Communications in Computer and Information Science, vol. 1307, pp. 540–563. Springer, Cham (2020). https://doi.org/10.1007/978-3-030-62833-8_40
53. Sánchez-Carrasco, C., Padilla, O., Toulkeridis, T.: Simulation of vehicle transit during an eventual eruption of the Cotopaxi volcano in the Valle de los Chillos, Central Ecuador. In: Rodriguez Morales, G., Fonseca, C.E.R., Salgado, J.P., Pérez-Gosende, P., Orellana Cordero, M., Berrezueta, S. (eds.) Information and Communication Technologies. TICEC 2020. Communications in Computer and Information Science, vol. 1307, 391–405. Springer, Cham. (2020). https://doi.org/10.1007/978-3-030-62833-8_29
54. Padilla-Almeida, O., León, P., Kirby, E.: Generation of a predictive model of urban growth through the application of cellular automata in the valley de los Chillos, Ecuador. In: Rodriguez Morales, G., Fonseca, C.E.R., Salgado, J.P., Pérez-Gosende, P., Orellana Cordero, M., Berrezueta, S. (eds.) Information and Communication Technologies. TICEC 2020. Communications in Computer and Information Science, vol. 1307, pp. 406–419. Springer, Cham (2020). https://doi.org/10.1007/978-3-030-62833-8_30
55. Sosa J., Puebla G., Morales M.: Plan de Desarrollo y Ordenamiento Territorial del Cantón Rumiñahui 2012–2025 (2012)
56. Bosque, J., Ramírez, L.: Localización de hospitales: analogías y diferencias del uso del modelo p-mediano en SIG raster y vectorial. Anales de Geografía de la Universidad Complutense **2001**, 53–82 (2001)

57. Sevtsuk, A., Mekonnen, M.: Urban network analysis. Revue internationale de géomatique–n, **287**, 305 (2012)
58. Álvarez, E., Bencomo, J., Esparza, M.: Consideraciones para la Optimización de Rutas en un Sistema de Transporte Escolar (STE). Cult. Científica y Tecnológica **12**(56), 15 (2015)
59. Barragán, K.T.: Análisis y gestión de flujos turísticos en el centro histórico de Málaga. Papers de Turisme **57**, 47–73 (2015)
60. López, R.: Movilidad cotidiana y accesibilidad: ser peatón en la ciudad de México (2014)
61. Miller, F.: GIS Tutorial for Marketing. ESRI Press, Redlands (2007)
62. Castellanos, L., López, M.: Estimación de rutas y tiempos de respuesta de los organismos de socorro en la ciudad de Manizales, apoyado en herramientas SIG. Universidad de Manizales, Proyecto de grado del programa de especialización en SIG (2015)
63. Cardozo, O., Gómez, P.M.: Teoría de Grafos y Sistemas de Información Geográfica aplicados al Transporte Público de Pasajeros en Resistencia (Argentina), pp. 89–111. Revista Transporte y Territorio N° 1, Universidad de Buenos Aires (2009)
64. Gallego, J., Labrador, E., Cabanillas, F.: Diseño de un modelo de asignación de viajes con aplicaciones SIG para la gestión de planes de movilidad urbana sostenibles en ciudades medias. GeoFocus. Revista Internacional de Ciencia y Tecnología de la Información Geográfica **13**(1), 1–21 (2013)
65. Sevtsuk, A., Mekonnen, M.: Urban network analysis. Revue internationale de géomatique– n **287**, 305 (2012)
66. Hernández, A.: Herramientas SIG para el estudio de la Carmona Romana = GIS tools to study (2012)
67. Dijkstra, E.: A note on two problems in connection with graphs. Numer. Math. **1**, 269–271 (1959)
68. Skiena, S.: Dijkstra's algorithm. In: Implementing Discrete Mathematics: Combinatorics and Graph Theory with Mathematica, pp. 225–227. Addison-Wesley, Reading (1990)
69. Zhan, F.B., Noon, C.E.: Shortest path algorithms: an evaluation using real road networks. Transp. Sci. **32**(1), 65–73 (1998)
70. Deng, Y., Chen, Y., Zhang, Y.: Fuzzy Dijkstra algorithm for shortest path problem under uncertain environment. Appl. Soft Comput. **12**(3), 1231–1237 (2012)
71. Johnsonbaugh, R.: Matemáticas Discretas. Pearson Prentice Hall, Upper Saddle River (2005)
72. Daskin, M.S.: Network and Discrete Location. Models, Algorithms and Applications. John Wiley & Sons, Inc., New York (1995)
73. Morgan, M.G., Fischhoff, B., Bostrom, A.: Risk Communication: A Mental Approach. Cambridge University Press, Cambridge (2012)
74. Visschers, V.H., Meertens, R.M., Passchier, W.W.: Probability information in risk communication: a review of the research literature. Risk Anal. **29**, 267–287 (2009)
75. Stein, S., Geller, R.J.: Communicating uncertainties in natural hazard forecasts. EOS Trans. Am. Geophys. Union **93**, 361–362 (2012)
76. Marti, J.: Scientific communication of uncertainty during volcanic emergencies. In: Loughlin, S.C., Sparks, R.S.J., Brown, S.K., Jenkins, S.F., Vye-Brown, C. (eds.) Global Volcanic Hazards and Risk, 371 pp. Cambridge University Press, Cambridge (2015)
77. Municipality of Rumiñahui: Contingency Plan of an Eventual Eruption of the Volcano Cotopaxi, 44 pp. Sangolquí, Ecuador (2015)
78. Lagos, M., Fritz, H.M.: Field survey of the 1 April 2014 Iquique tsu-nami along the coasts of Chile and Peru. In: AGU Fall Meeting Abstracts, vol. 1, p. 4398. December 2014

Smart City Planning Based on Landslide Susceptibility Mapping Using Fuzzy Logic and Multi-criteria Evaluation Techniques in the City of Quito, Ecuador

Daniela Salcedo, Oswaldo Padilla Almeida, Byron Morales, and Theofilos Toulkeridis(✉)

Universidad de las Fuerzas Armadas ESPE, Sangolquí, Ecuador
ttoulkeridis@espe.edu.ec

Abstract. Landslides are the most recurrent natural hazards in the Metropolitan District of Quito (DMQ), affecting sometimes lives but extremely frequently and severely the traffic and associated infrastructure. The present research proposes the calculation of the landslide susceptibility cartographic model in the city of Quito and its main highway, the Simón Bolívar avenue, using the Fuzzy logic and multicriteria evaluation techniques in geographic information systems (GIS). Based on the "Today and the past son key to the future" principle, landslides have been located using aerial photographs and field work. Based on the characteristics of historical landslides, photointerpreted and previous studies, the causal factors have been variable such as topography, structural geology, lithology, precipitation, water network, vegetation cover, among others. Each factor has been processed, analyzed and standardized according to its relationship to the occurrence of landslides, by means of a sinusoidal linked function that assigns to each element a degree of correlation [0, 1] to the diffuse set. The landslide vulnerability map has been obtained from the combination of causal factors by map algebra, such as weighting techniques that include the hierarchical analysis process (HAP) and the weighted linear line (WLL), whose validation considered the locations of inventoried landslides.

Keywords: Landslide susceptibility map · Fuzzy logic · GIS · Simón Bolívar highway · Quito

1 Introduction

Landslides constitute one of the most damaging natural hazards in mountainous areas generating human, material and economic losses [1]. Landslides and other types of mass movement are responsible for 17% of all natural hazard fatalities worldwide [2], while economic and social losses due to such types of catastrophic events are able to be reduced by a more efficient land use and projection [3]. Landslide risk (R) has been usually described as a function of the probability and consequence of landslides [4]

$$R = \int(P, C_i) \tag{1}$$

S. Berrezueta and K. Abad (Eds.): *Doctoral Symposium on Information and Communication Technologies - DSICT*, LNEE 846, pp. 89–103, 2022.
https://doi.org/10.1007/978-3-030-93718-8_8

Where P is the probability of risk, and C is a vector C_i for all potential consequences. Quantifying potential consequences C_i such as loss of life or damage is difficult because of incomplete or limited historical records that affect population and infrastructure estimates at risk [5]. However, the probability of risk *(P)* has been analyzed from different evaluation approaches such as qualitative, through geological and geomorphological analysis [6] and such as quantitative, considering causal factors and probabilistic methods, the same ones that have taken force due to computational and geospatial technological improvements [7]. However, the semi-quantitative have been through cause, effect and consequence where instability factors have been selected and weighted according to their relationship with the presence of the event and the occurrence of losses or damages. This has been previously applied within the risk classification system for cut and refill of slopes as well as for natural slopes in future development zones [8].

The susceptibility maps of landslides have been developed through several methods, of which the most used are the inventory of landslides based on probabilistic techniques, the deterministic as well as the heuristic and statistical techniques [9]. The first three methods assign weights to a series of causal factors used according to the researcher's experience, while statistical analysis considers causal factors in areas with similar environmental conditions to those of past landslides [10]. Nowadays, non-parametric techniques such as cellular automata, Fuzzy logic and artificial neural networks are applied [11–13].

Geographic Information Technology such as Remote Sensors (RS), Geographic Information Systems (GIS) and Global Positioning Systems (GPS) allow the management, assessment and identification of landslide risk. The use of GIS has been positioned because of its ease of collection and analysis of spatial data [14]. GIS supported by multicriteria decision analysis, such as: weighted average, analytical hierarchy process, weighted linear combination, order among others, are useful procedures for the recognition, evaluation and prognostication of risk by landslides [15]. The most appropriate approach and methods for the evaluation of susceptibility and hazards to landslides have not been defined [16], their selection depend on the characteristics of the study area both their baseline, historical, developed research and work scale [17].

The risk assessment for landslides by means of cartography allows communication with direct involved persons for future impacts and public authorities responsible for risk management and prevention [18], allowing the development of effective strategies for stabilization of vulnerable areas, urban planning and implementation of infrastructure [19]. According to [20], 90% of landslide losses could be avoided if the problem would have been previously recognized prior the catastrophic event [21].

Therefore, the main aim of this study has been to generate a cartographic model of susceptibility to landslides of the city of Quito in general and its main highway named Simón Bolívar Avenue, using Fuzzy Logic methodology and multicriteria evaluation in GIS adjusted to historical records that allows the preventive evaluation of susceptible zones in order to avoid human and financial losses, as well as of infrastructure.

2 Study Area

The study area evaluates the susceptibility of the main axis of the Simón Bolívar Avenue, Quito's main highway from its southern end (Turubamba) along with its connection with the Bicentennial Avenue towards its northern end at the Pan-American Highway. In order to validate the results with historical information, we considered an area of indirect influence (AII) with an extension of approximate 393.3 Km^2, which included the zonal administrations of Los Chillos, Quitumbe, Eloy Alfaro, Manuela Sáenz, Tumbaco, Eugenio Espejo, Calderón and La Delicia, while the area of direct influence (ADI) comprised of some 22.53 Km^2 covering a radius of 275 m on each side of the main road axis (Fig. 1). The Simón Bolívar Avenue is the main highway of the city of Quito and forms part of its eastern peripheral ring. This highway has been conceived within the mobility plan as a traffic solution when connecting with the northern and southern sectors as well as the valleys bordering the Metropolitan District of Quito.

Within the ADI, the volcano - sedimentary fill comprises deposits of lava, agglomerates, tuffs, volcanic ash, layers of pumice and undifferentiated sediments being part of the geological formations such Machángara (Pleistocene) and Chiche as well as primary and colluvial volcanic deposits that have been generated during the Holocene [22]. The topography varies between 1915 and 3380 m above sea level (m.a.s.l.). The Machángara river crosses the road artery in the sector of Guápulo, which raised from the tributaries: Ortega, Shanshayacu, Rio Grande and Capulí or Machángara, conformed by 22.5 km and bordered by very steep slopes, is the main stream and more (EPMAPS), along with the Monjas River of the DMQ, where 76.3% of urban wastewater flows from the south, center and north of the city to the level of the Guayasamín tunnel without any type of previous treatment [23], having an organic load with 80% of domestic and 20% industrial origin, respectively [24]. Soil is commonly used in infrastructure and the vegetation cover is made up of both wet and dry shrubs and broadleaf cultivated vegetation [25]. The maximum precipitation is given to the southern sector of the city and decreasing towards the north [25].

Simón Bolívar Avenue is the first roadway with the highest vehicular traffic of the DMQ. Through there, 74,469 vehicles pass through [26]. However, there are six areas prone to landslides in the winter season [27]. Its implementation in mountainous area, proximity to natural channels, deficient or absent channeling of rainwater, human settlements, anti-technical discharges of sewage and deforestation have influenced the frequency of landslides especially in the rainy winter season where their constant work of soil removal and maintenance represent a high public investment in rehabilitation and reconstruction, incurring high levels of congestion, human losses, and material, affecting infrastructure and commercial sectors [28].

In 2011, the road has been completely disabled for two weeks due to the landslide of 49,700 m^3 of land that registered 5 fatalities and 120 evacuated in the sector of Forest IV [29], while the Municipality invested 636,478 dollars in stabilization works [30]. Considering the vulnerability of the road, the Metropolitan Public Company for Mobility and Public Works (EPMMOP) has taken precautionary measures through works of cleaning both the vegetation layer and ditches of coronation and tracks, slope profiling and implementation of complementary works for water eviction, in sectors such as El Madrigal, San Isidro de Puengasí, Chaquishahuaycu ravine, among others, where a variety of slope stabilization works have been performed.

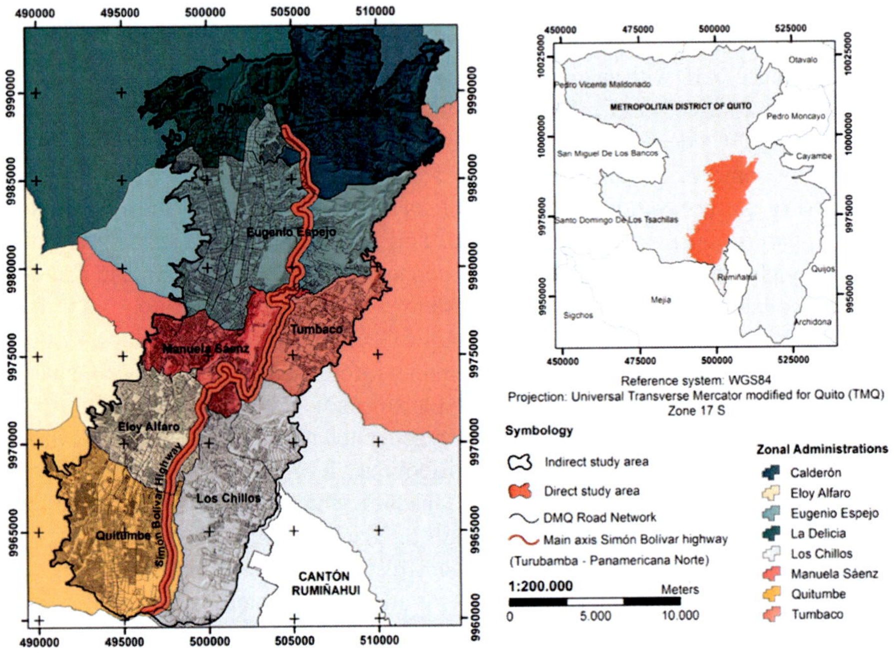

Fig. 1. Location map of the study area

In 2016, EPMMOP handled 28 landslides along Avenida Simón Bolívar, of which 11 have been in Guápulo and the remaining in the sectors Alma Lojana, Santa Rosa, Monteolivo, Las Bromelias, Miravalle 4, Oswaldo Guayasamín, Chaquishahuaycu Creek, Zámbiza, Algeria Alta, El Troje, Llano Chico, Granados, Guajaló, San Francisco, La Forestal and Monjas. In 2017, the DMQ recorded the highest precipitation levels of the last 30 years [30], where main roads such as Av. Velasco Ibarra and Av. Simón Bolívar were heavily affected. Between October 2016 and May of 2017, Quito attended 965 emergencies, of which 443 have been landslides [31]. 14.7 million dollars have been allocated to the prevention and response plan for rainy season, plus further two million US Dollars from the emergency fund, where 1,870 officials have been mobilized for disaster prevention and care [30], a management that has been recurrent in previous years. The city itself can be reached by a variety of natural hazards [50–56].

3 Materials and Methods

The areas prone to landslides have been determined using a heuristic method, a qualitative analysis where a series of factors related to the occurrence of landslides is selected, hierarchized and weighed. The susceptibility of these landslides has been solved through decision tools applied in GIS [32–34] due to the ease of handling spatial information, versatility of representation and quality of results. The applied methodology has been based on the principle "Today and the past are keys to the future", considering that: (1) Future landslides will have the same geological, geomorphological and water factors as those already in the past or in nearby previously studied areas; and (2) Causal factors will be represented by spatial data contained in a GIS database that will allow diffuse overlap [35]. The causal factors have been standardized using a membership function according to the relationship between the dependent variable (presence or absence of landslides) and independent variables (landslides/causal factors). The obtained results from each map factor have been weighted by hierarchical analysis (AHP) according to their relevancy with the occurrence of landslides.

3.1 Determination of Causal Factors

The causal or instability factors have been divided into conditioning factors and triggers, the former depending on the characteristics of the area such as topography, geology, lithology, soil type, drainage and vegetation cover [19]. The second increase the probability of occurrence, being of natural origin such as earthquakes, precipitation intensity and natural or anthropic erosion due to land use and construction operations [36]. Both of them vary significantly from one region to another [21], so it has been fundamental to determine them by means of a retrospective analysis. According to geomorphological factors, slope, climate and human intervention play an important role [37]. According to the National Institute of Statistics and Censuses (INEC), rapid urbanization within the DMQ has increased settlements in inadequate areas such as slopes or edges of rivers and streams that produce vulnerable sectors especially in winter [25, 38]. At the same time the implementation of roads in mountainous regions has generated problems of instability that geologists and civil engineers have been aware of [39, 40].

The scarce vegetation cover of steep slopes has been affected by forest fires, being one of the most frequent phenomena of the DMQ, where approximately nine events on the Simón Bolívar Avenue and surrounding areas have been registered in 2016 [41]. Considering the historical record of landslides on Avenida Simón Bolívar in sectors such as Forestal IV (2011), Buenos Aires (2015), detour of Granados (2016 and 2017), El Troje (2017), Monteolivo (2017), Puengasí (2017), La Forestal (2017), as well as causal factors exposed in studies developed by the Municipality, projects in the Monjas - Ferroviaria - La Magdalena - Itchimbia area and the zonal administration Eugenio Espejo [33, 42]. Generating or causal factors have been considered to be the topography, geology, lithology, vegetation cover, precipitation, water network and the main accessibility.

3.2 Standardization, Weighting and Evaluation

As each factor has a different scale of measurement and cartographic representation (vector or raster) to perform a multicriteria analysis, therefore the input layers need to be standardized from their original values to values in a range of 0 to 1 according to the normalization formula (Eq. (2)). The standardization has been performed by Fuzzy logic in order to assign to each element a membership value according to the degree of membership in the set. The membership function is of sinusoidal type whose value oscillates between 0 and 1 by the use of sine curves (Fig. 2) and cosine (Fig. 3) in a range of 0 to 90° ($\pi/2$ radians), its occupation will depend on the relationship between the analyzed variable and the probability of landslides [33]. The limits of the function correspond to the maximum and minimum values of each factor taking into account that all values represent a level of risk.

Normalization:

$$N = \frac{V_o - V_{min}}{V_{max} - V_{min}} \tag{2}$$

Where: V_0: Original value; V_{min}: Minimum value and, V_{max}: Maximum Value. Directly proportional (2nd Case):

$$\mu_A(V_o) = \sin\left(\frac{\pi}{2} \times \frac{V_o - V_{min}}{V_{max} - V_{min}}\right) 0 \leq \mu_A(V_o) \leq 1 \tag{3}$$

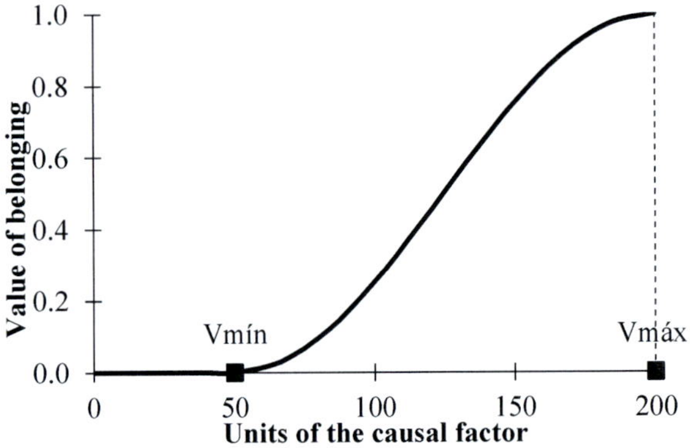

Fig. 2. Sine curve belonging function from 0 to π/2 radians [33]

Inverse proportional (3rd Case):

$$\mu_A(V_o) = \cos\left(\frac{\pi}{2} \times \frac{V_o - V_{min}}{V_{max} - V_{min}}\right) \quad 0 \leq \mu_A(V_o) \leq 1 \tag{4}$$

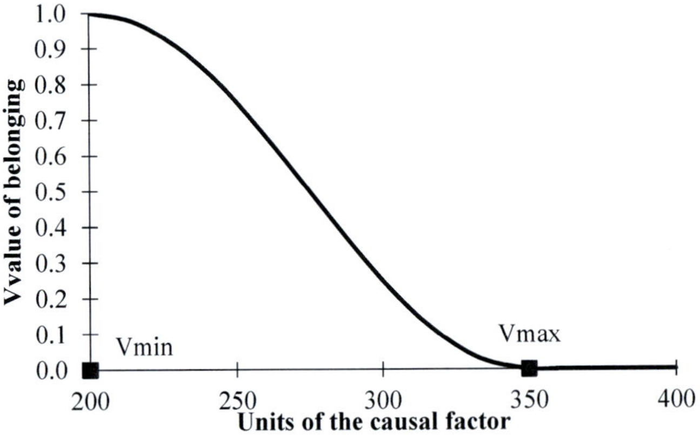

Fig. 3. Curve belonging property Cosine range from 0 to π/2 radians [33]

Where $\mu_A(V_o) = 1 V_o$ belongs entirely to the set, $\mu_A(V_o) = 0$ si V_o does not belong to the set and $0 < \mu_A(V_o) < 1$ if V_o is partially in the set.

4 Results and Discussion

The landslides have been located and inventoried through aerial photographs taken in 2011 by the IGM, Digital Terrain Model and Ortho-mosaic scale 1:1000 DMQ, provided by the Ministry of agriculture, livestock, aquaculture and fisheries based on the Project National System of Information and Management of Rural Lands and Technological Infrastructure (SIGTIERRAS), considering areas of rupture, bare soil and other typical geomorphological characteristics. The map of the sliding inventory has been used to evaluate the spatial distribution of landslides and to determine their triggering factors in the study area. A total of 45 landslides have been identified at Avenida Simón Bolívar, through photointerpretation and field tests with the use of a Global Positioning System (GPS) with a precision of 4 m.

4.1 Slope Map

The slope map has been generated from the scale curves 1:5000 and MDT. Altitudes vary from 1915 m to 3380 m, where the maximum slope has been of 88.68°, and the minimum corresponds to flat areas with 0°.

4.2 Vegetation Cover Map

The vegetation cover map has been weighted considering similar studies within DMQ [33, 43], considering a susceptibility range from 1 (Low) to 5 (Very High).

4.3 Precipitation Map

The slope map has been generated from the scale curves 1:5000 and MDT. Altitudes vary from 1915 m to 3380 m, where the maximum slope has been of 88.68°.

The annual precipitation records for 25 years (1991–2015) have been provided by the National Institute of Meteorology and Hydrology (INAMHI), whose nearby stations have been listed in [57]. The estimation of precipitation in stations with incomplete statistics have been performed using the average ratio completion method or the normal ratio (Eq. (5)), recommended for mountain areas where the precipitation of nearby stations generally differ by more than 10%, by which the relationship between the station "x" and its nearby stations in a period has been considered [44].

$$P_x = \frac{1}{n}\left[\left(\frac{N_x}{N_1}\right)P_1 + \left(\frac{N_x}{N_2}\right)P_2 + \ldots + \left(\frac{N_x}{N_n}\right)P_n\right] \tag{5}$$

Where n represents number of rainfall stations with continuous recorded data close to the station "x", which will be completed in its registry; Px represents precipitation of the station "x" during the time period to be completed; P_1 to P_0 shall be the precipitation of nearby stations during the time period to be completed; Nx represents the annual average season precipitation of station "x"; N_1 to N_{n} represents the annual average precipitation of the stations nearby [45].

The homogeneity of the data has been estimated by the average ratio completion method or the normal ratio being verified by double mass analysis", which consisted of constructing a cumulative double curve that relates the cumulative annual precipitation totals of the station "*x*" and its near station [45]. The obtained correlation coefficients from the double mass curve for all the stations have been close to one, which evidenced the homogeneity of the data. The results have been verified by means of test of streaks, where their homogeneity has been confirmed.

4.4 Lithology Map

The lithology map has been weighted by a numerical scale where 1 represents the highest vulnerability and 6 the lowest, according to field experience. In flat areas, the effect of lithology has been reduced due to the absence of landslide vulnerability.

4.5 Other Maps

The geological, hydric and road parameters have been analyzed by proximity using the "Euclidean Distance" tool (ArcGIS), distance from each cell in the raster to the nearest source (geological fault, river and road) (Figs. 4, 5, 6 and 7).

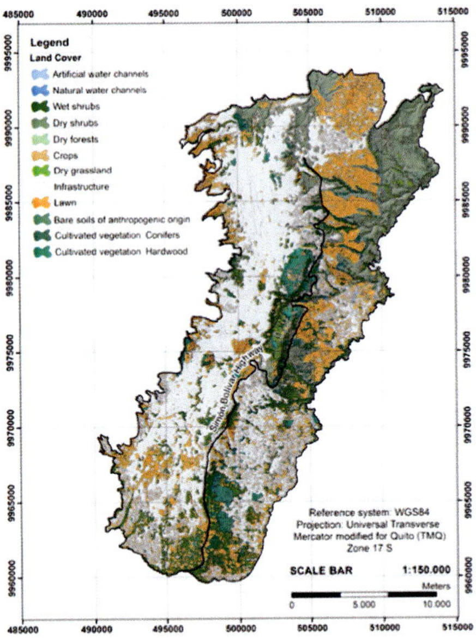

Fig. 4. Map of land cover

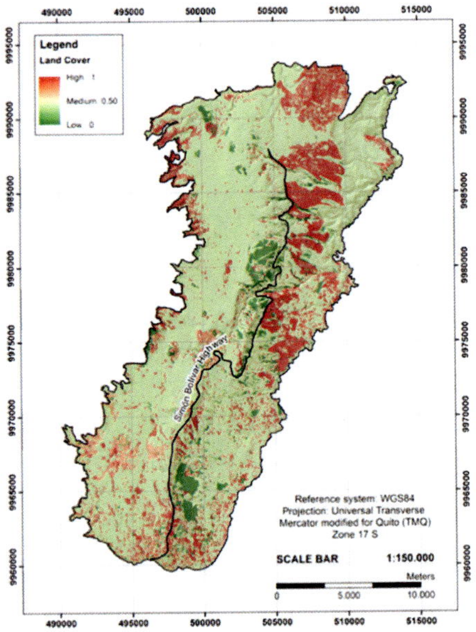

Fig. 5. Normalized land cover

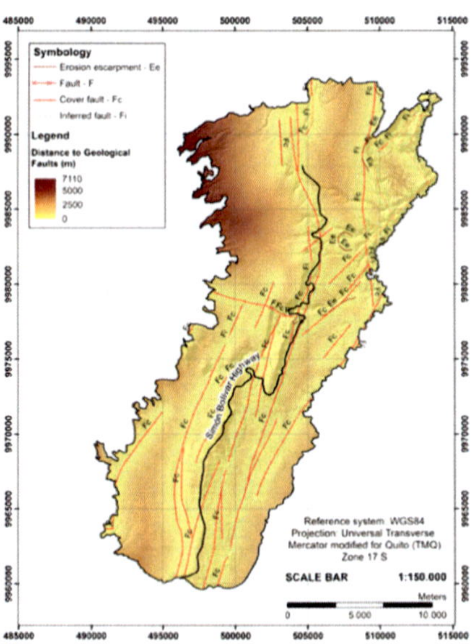

Fig. 6. Distance of geological faults.

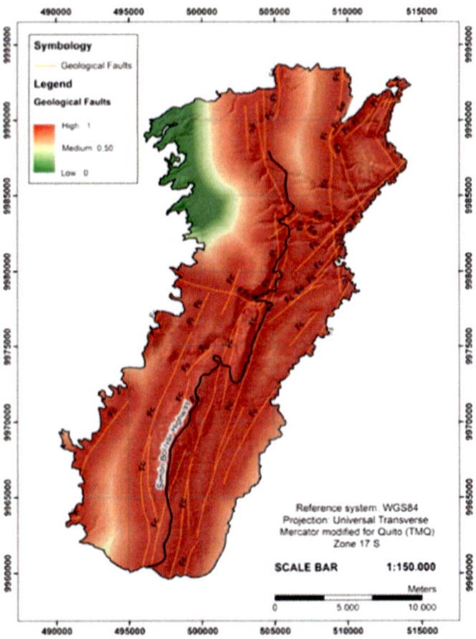

Fig. 7. Map with normalized distance of geological faults

4.6 Map of Susceptibility to Landslides Through the Hierarchical Analysis Method (AHP)

The geological, hydric and road parameters have been analyzed by proximity using the "Euclidean Distance" tool (ArcGIS), distance from each cell in the raster to the nearest source (geological fault, river and road).

The matrix of AHP considers three principles: decomposition, comparative judgment and synthesis of priorities [46], through which the problem (and its weights) is decomposed in a hierarchical structure, whose weighting depends on the comparison of criteria by pairs and the multiplication between the respective hierarchical levels [8]. Assigned values range from ("much less important than"); 1 ("equally important than") and, 9 ("much more important than") [47, 48] (Table 1).

Weighted Linear Sum

The susceptibility map has been the result of the combination of causal factors with their respective weights obtained by AHP as shown in Eq. (6)

$$I = \sum_{j=1}^{n} W_j \times x_{ij} \quad (6)$$

Where: I: Susceptibility index, W_j: factor weight j, x_{ij}: normalized value of each map and n: number of factors. Resulting in:

$$f_{(x)} = 0{,}379x_1 + 0{,}246x_2 + 0{,}159x_3 + 0{,}102x_4 + 0{,}054x_5 + 0{,}036x_6 + 0{,}024x_7 \quad (7)$$

Table 1. Hierarchical analysis of causal factors (*CR: 0,030)

	(1)	(2)	(3)	(4)	(5)	(6)	(7)	Weigh-ting
(1) Slope	1	3	9	2	8	7	6	0,379
(2) Precipitation	1/3	1	7	1/2	6	3	2	0,159
(3) Plant cover	1/9	1/7	1	1/8	1/2	1/3	1/6	0,024
(4) Lithology	1/2	2	8	1	7	6	3	0,246
(5) Road distance	1/8	1/6	2	1/7	1	1/2	1/3	0,036
(6) River distance	1/7	1/3	3	1/6	2	1	1/3	0,054
(7) Geol. fault distance	1/6	1/2	6	1/3	3	3	1	0,102

Where: x_1: Pending; x_2: Lithology; x_3: Precipitation; x_4: Distance to geological faults; x_5: Distance to rivers; x_6: Distance to roads y, x_7: Vegetation cover.

Model Adjustment

$$N = Value_{Measured} - Value_{Calculated} \tag{8}$$

$$N = 1 - Y$$

$$N = 1 - 0,1620 \tag{9}$$

$$N = 0{,}838$$

Where: N: *Adjustment*; $Value_{Measured}$: 1 or value of the sample points and, $Value_{Calculated}$: standard deviation of the resulting model.

Thus, the level of slip irrigation has been classified according to [49], based on the model of Fuzzy logic obtained in a range of 0 to 1 [35] and represented in color scales (Table 2).

The obtained susceptibility map from the combination of Fuzzy methodology and multicriteria evaluation yielded a standard error of 0.1620 with an adjustment value of 83.8%. Where 2% of AEI has critical susceptibility, 10% high, 37% average, 49% low and 2% nil. Within the AED, 5% have critical susceptibility, 19% high, 58% average and 18% low. The obtained results have been in accordance with the spatial distribution of landslides, together with historical records in areas such as La Ferroviaria, Guápulo, Oswaldo Guayasamín, Granados Connection, among others (Fig. 8). The new vulnerability map has a more detailed aspect of all used parameters when compared with the existing map of the municipality of Quito.

Table 2. Slip susceptibility zonation according to the diffuse output membership functions.

Zone	Fuzzy belonging function	Description level of susceptibility	Color of scales
I	<0.1	None	Dark green
II	–0.40	Low	Light green
III	0.40–0.60	Medium	Yellow
IV	0.60–0.75	High	Orange
V	>0.75	Critical	Red

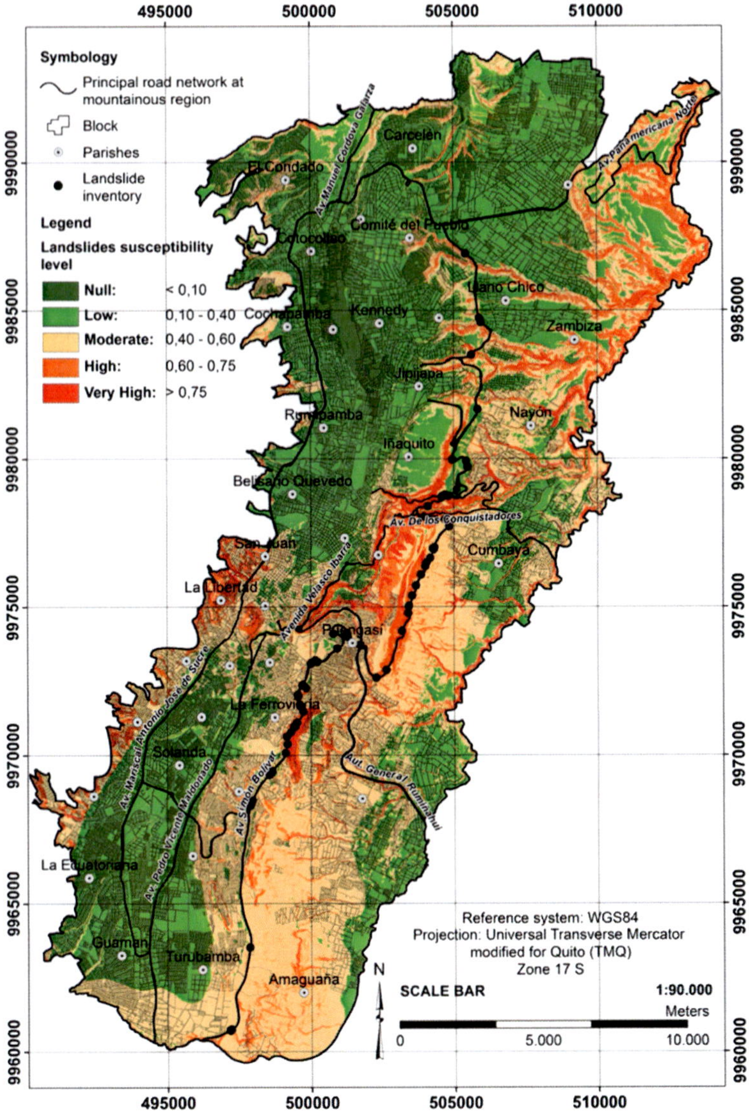

Fig. 8. Map of susceptibility to landslides derived from the AHP method

5 Conclusions

The Simón Bolívar Avenue has a medium to high level of susceptibility to landslides, where the most critical sectors are located in the parishes such as La Ferroviaria, Puengasí and Itchimbía. A preventive analysis of the area certainly will be more economical than the implementation of corrective measures once the event has been triggered.

Topography and geology have been the factors that mainly defined the level of susceptibility as they condition the effect of detonators such as precipitation and anthropic intervention.

The parameters and methodology selected for the study area have been considered adequate and coincide with the factors established in the field.

The cartographic models of susceptibility to landslides have been useful tools for urban planning and implementation of strategic infrastructure such as road corridors and allow to locate new vulnerable areas beyond the historically recognized. Simuyltaneously they constitute the baseline for developing models for geotechnical and seismic zoning within the DMQ.

The fuzzy methodology combined with multicriteria analysis would have a better correlation with historical data and the spatial distribution of inventoried slides, if the information base of each factor provided by the competent entities would have been updated, validated and at scales of greater precision.

Social parameters such as land use, evolution of the urban spot in illegally constituted areas and implementation of strategic infrastructure at the edge of slopes should be included within the causal factors. The inventory of resources affected and at potential risk through multitemporal analysis would make it easier to estimate the human and economic losses that the landslides have generated and could affect the public and private sector.

References

1. Álcantara-Ayala, I., Esteban-Chavez, O., Parrot, J.: Landsliding related to land-cover change: A diachronic analysis of hillslope instability distribution in the Sierra Norte, Puebla Mexico. CATnENA **65**, 152–165 (2006)
2. Lacasse, S., Nadim, F., Kalsnes, B.: Living with landslide risk. Geotech. Eng. J. SEAGS AGSSEA **41**(4), 13 (2010)
3. Rajakumar, R., Sanjeevi, S., Jayaseelan, S.: Landslide susceptibility mapping in a hilly terrain using remote sensing and GIS. J. Indian Soc. Remote Sens. **35**, 31–42 (2007)
4. Andersson-Sköld, Y., Bergman, R., Johansson, M.: Landslide risk management—a brief overview and example from Sweden of current situation and climate change. Int. J. Disast. Risk Reduct. **3**, 44–61 (2013)
5. Grahn, T., Jaldell, H.: Assessment of data availability for the development of landslide fatality curves. Landslides **14**(3), 1113–1126 (2016). https://doi.org/10.1007/s10346-016-0775-6
6. Park, S., Choi, C., Kim, B.: Landslide susceptibility mapping using frequency ratio, analytic hierarchy process, logistic regression, and artificial neural network methods at the Inje area Korea. Environ. Earth Sci. **68**, 1443–1464 (2013)
7. Bai, S.-B., Wang, J., Lü, G.-N.: GIS-Based logistic regression for landslide susceptibility mapping of the Zhongxian segment in the three Gorges area. China. Geomorphol. **115**, 23–31 (2010)

8. Castellanos Abella, E., Van Westen, C.J.: Generation of a landslide risk index map for Cuba using spatial multi - criteria evaluation. Landslides **4**, 311–325 (2007)
9. Chiessi, V., Toti, S., Vitale, V.: Landslide susceptibility assessment using conditional analysis and rare events logistics regression: a case - study in Antrodoco Area (Rieti, Italy). J. Geosci. Environ. Protect. **4**(12), 1–21 (2016)
10. Isik, Y.: Landslide susceptibility mapping using frequency ratio, logistic regression, artificial neural networks and their comparison: a case study from Kat landslides (Tokat—Turkey). Comput. Geosci. **35**, 1125–1138 (2009)
11. Falaschi, F., Giacomelli, F., Federici, P.: Logistic regression versus artificial neural networks: Landslide susceptibility evaluation in a sample area of the Serchio River valley. Italy. Nat. Hazards **50**, 551–569 (2009)
12. Pradhan, B., Lee, S.: Landslide susceptibility assessment and factor effect analysis: back-propagation artificial neural networks and their comparison with frequency ratio and bivariate logistic regression modelling. Environ. Model. Softw. **25**, 747–759 (2010)
13. Saro, L., Woo, J.S., Kwan-Young, O.: The spatial prediction of landslide susceptibility applying artificial neural network and logistic regression models: a case study of Inje. Korea. Open Geosci. **8**, 117–132 (2016)
14. Bayes, A.: Landslide susceptibility mapping using multi-criteria evaluation techniques in Chittagong Metropolitan Area Bangladesh. Landslides **12**, 1077–1095 (2015)
15. Feizizadeh, B., Blaschke, T.: GIS-multicriteria decision analysis for landslide susceptibility mapping: comparing three methods for the Urmia lake basin. Iran. Nat. Hazards **65**(3), 2105–2128 (2013)
16. Brabb, E.: Innovative approaches to landslide hazard and risk mapping. In: 4th International Symposium on Landslides, pp. 307–323 (1984)
17. Hervás, J., Barredo, J., Lomoschitz, A.: Elaboración de mapas de susceptibilidad de deslizamientos mediante SIG, pp. 169–180. Teledetección y Metódos de evaluación multicriterio.Aplicación a la depresión de Tirajana (Gran Canaria). Instituto Geológico y Minero de España (2012)
18. Andersson-Skold, Y., Falemo, S., Trembaly, M.: Development of methodology for quantitative landslide risk assessment-Example Gota river valley. Nat. Sci. **6**, 130–143 (2014)
19. Dahal, B.K., Dahal, R.K.: Landslide hazard map: tool for optimization of low-cost mitigation. Geoenviron. Disast. **4**(1), 1–9 (2017). https://doi.org/10.1186/s40677-017-0071-3
20. Brabb, E.: Proposal for worldwide landslide hazard maps. In: Proceedings of 7th International Conference and Field Workshop on Landslide in Czech and Slovak Republics, pp. 15–27 (1993)
21. Pardeshi, S.D., Autade, S.E., Pardeshi, S.S.: Landslide hazard assessment: recent trends and techniques. Springerplus **2**(1), 1–11 (2013). https://doi.org/10.1186/2193-1801-2-523
22. INIGEMM (Instituto Nacional de Investigación Geológico Minero Metalúrgico) (1978–1980). Geologic maps in a scale of 1:50000 of: Quito (1978), Sangolquí (1980) and El Quinche (1980). http://www.geoinvestigacion.gob.ec/mapas-tematicos-1-50-000/
23. Moreano, M.: La Biografía Secreta De Las Aguas Quiteñas. Ecuador terra incognita **65**, 10–19 (2010)
24. Reinoso Chisaguano, I.C.: Evaluación Ambiental Del Río Machángara. Unpubl. Thesis Escuela Politécnica Nacional, Quito, Ecuador, 178pp (2015)
25. Secretaría de Seguridad y Gobernabilidad del DMQ: Atlas De Amenazas Naturales y Exposición De Infraestructura Del Distrito Metropolitano De Quito (vol. Segunda Edición). Quito, Ecuador, 126pp (2015)
26. El Comercio: La Mariscal Sucre, Segundo Puesto En El 'Top Ten' De Las Más Transitadas De Quito. http://www.elcomercio.com/actualidad/quito-transito-avmariscalsucre-congestion.html (2015)

27. El Telégrafo: La EPMMOP Estabilizará Dos Taludes En Av. Simón Bolívar. http://www.eltelegrafo.com.ec/noticias/quito/11/la-epmmop-estabilizara-dos-taludes-en-av-simon-bolivar (2013)
28. Toulkeridis, T., Rodríguez, F., Arias Jiménez, N.: Causes and consequences of the sinkhole at El Trébol of Quito, Ecuador – implications for economic damage and risk assessment. Nat. Hazards Earth Syst. Sci. **16**, 2031–2041 (2016)
29. El Comercio: Estudios En La Forestal No Terminan (2011). http://www.elcomercio.com/actualidad/quito/estudios-forestal-no-terminan.html
30. Prensa Alcaldía de Quito: Municipio Atiende Emergencias Por Deslizamientos (2017). http://prensa.quito.gob.ec/index.php?module=Noticias&func=news_user_view&id=25751&umt=Municipio%20atiende%20emergencias%20por%20deslizamientos
31. El Universo: Municipio De Quito Destinó Más De 15 Millones De Dólares Para La Época Invernal (2017). http://www.eluniverso.com/noticias/2017/06/11/nota/6227731/municipio-quito-destino-mas-15-millones-dolares-epoca-invernal
32. Ciampalini, A., Raspini, F., Bianchini, S.: A landslide susceptibility map of the Messina Province (Sicily, Italy). Landslides and engineered slopes. Exp. Theory Pract. **1**, 657–661 (2016)
33. Zafrir Vallejo, R., Padilla Almeida, O., Cruz D'Howitt, M.: Numerical probability modeling of past, present and future landslide occurrences in Northern Quito, Ecuador – economic implications and risk assessment". In: 5th International Conference on eDemocracy and eGovernment, ICEDEG 2018, pp. 117–125 (2018)
34. Palacios Orejuela, I., Toulkeridis, T.: Evaluation of the susceptibility to landslides through diffuse logic and analytical hierarchy process (AHP) between Macas and Riobamba in Central Ecuador. In: 7th International Conference on eDemocracy and eGovernment, ICEDEG 2020, pp. 201–207 (2020)
35. Maryam, I.: A comparative study of fuzzy logic approach for landslide susceptibility mapping using GIS: an experience of Karaj dam basin in Iran. Procedia Soc. Behav. Sci. **19**, 668–676 (2011)
36. Singh, A.K.: Bioengineering techniques of slope stabilization and landslide mitigation. Disast. Prev. Manag. Int. J. **19**, 384–397 (2010)
37. Biju Abraham, P., Shaji, E.: Landslide hazard zonation in and around Thodupuzha-Idukki-Munnar road, Idukki district, Kerala: a geospatial approach. J. Geol. Soc. India **82**(6), 649–656 (2013). https://doi.org/10.1007/s12594-013-0203-7
38. Sepúlveda, S., Petley, D.: Regional trends and controlling factors of fatal landslides in Latin America and the Caribbean. Nat. Hazard Earth Syst. Sci. **15**(8), 1821–1833 (2015)
39. Silvers, K., Griffiths, J.: Landslide Affecting the Vía Interoceánica, IAEG2006 Paper number 234: 8pp. East of Quito, Ecuador (2006)
40. Fookes, P., Sweeney, M., Manby, C.: Geological and geotechnical engineering aspects of low-cost roads in mountainous terrain. Eng. Geol. **21**, 1–152 (1985)
41. Cantuña, J.G., Bastidas, D., Solórzano, S.: Design and implementation of a wireless sensor network to detect forest fires. In: 2017 4th International Conference on eDemocracy and eGovernment, ICEDEG 201729 June 2017, Article number 7962508, pp. 15–21 (2017)
42. Jaramillo Castelo, C.A., Padilla, O., Cruz D´Howitt, M.: Comparative determination of the probability of landslide occurrences and susceptibility in central Quito, Ecuador. In: 5th International Conference on eDemocracy and eGovernment, ICEDEG 2018, pp. 136–143 (2018)
43. Loarte Merino, G.V. Determinación de zonas susceptibles a movimientos en masa por factores condicionantes y desencadenantes en la parroquia Lloa, del cantón Quito en la provincia de Pichincha. Unpublished thesis. 100 pp. Universidad San Francisco de Quito, Cumbaya, Ecuador (2013)

44. UNESCO – ROSTALC: Guía Metodológica Para La Elaboración Del Balance Hídrico Para América Del Sur. Montevideo, Uruguay: Oficina Regional de Ciencia y Tecnología de la UNESCO para América Latina y el Caribe (1982). http://www.unesco.org.uy/phi/biblioteca/files/original/4028d09481fd2531c30f5cd6660298c3.pdf
45. Monsalve, G.: Hidrología En La Ingeniería, 2da edn. 358pp. Alfaomega, Colombia (2009)
46. Malczewski, J.: A GIS-based approach to multiple criteria group decision-making. Int. J. Geogr. Inf. Syst. **10**, 955–971 (1996)
47. Eastman, J.R.: The IDRISI Selva Tutorial Manual Version 17, 354pp. Clark Labs, Clark University (2012)
48. Aviles-Campoverde, D., Chunga, K., Ortiz-Hernández, E.: Seismically induced soil liquefaction and geological conditions in the city of Jama due to the Mw7.8 Pedernales earthquake in 2016, NW Ecuador. Geosciences **11**, 20 (2021)
49. Anbalagan, R.: Landslide hazard evaluation and zonation mapping in mountainous terrain. Eng. Geol **32**(4), 269–277 (1992)
50. Rodriguez, F., Toulkeridis, T., Padilla, O., Mato, F.: Economic risk assessment of Cotopaxi volcano Ecuador in case of a future lahar emplacement. Nat. Hazards **85**(1), 605–618 (2017)
51. Toulkeridis, T., Zach, I.: Wind directions of volcanic ash-charged clouds in Ecuador – Implications for the public and flight safety. Geom. Nat. Hazards Risks **8**(2), 242–256 (2017)
52. Toulkeridis, T.: The Evaluation of unexpected results of a seismic hazard applied to a modern hydroelectric center in central Ecuador. J. Struct. Eng. **43**(4), 373–380 (2016)
53. Toulkeridis, T., et al.: Causes and consequences of the sinkhole at El Trébol of Quito, Ecuador - Implications for economic damage and risk assessment. Nat. Hazards Earth Sci. Syst. **16,** 2031–2041 (2016)
54. Toulkeridis, T., et al.: Two independent real-time precursors of the 7.8 Mw earthquake in Ecuador based on radioactive and geodetic processes—powerful tools for an early warning system. J. Geodyn. **126,** 12–22 (2019)
55. Segura-Alcívar, M., Rodriguez-Espinoza, F., Toulkeridis, T.: Potential risk analysis of fuel storages in central Quito, Ecuador. In: Proceedings of the International Conference on Natural Hazards and Infrastructure 2019, 2nd International Conference on Natural Hazards and Infrastructure, ICONHIC 2019; Chania; Greece; 23 June 2019 through 26 June 2019; Code 257429 (2019)
56. Toulkeridis, T., et al.: A potential early warning system for earthquakes based on two independent real-time precursors – the case of Ecuador's 7.8 Mw in 2016. In: Proceedings of the International Conference on Natural Hazards and Infrastructure 2019, 2nd International Conference on Natural Hazards and Infrastructure, ICONHIC 2019; Chania; Greece; 23 June 2019 through 26 June 2019; Code 257429 (2019)
57. INAMHI. Anuario meteorológico. Instituto Nacional de Meteorología e Hidrología (2017). http://www.inamhi.gob.ec/

Free Data Processing Applied to Detect Changes in Land Use Coverage at Biodiversity Hotspots of the Amazon

Marco Heredia-R[1,2,3](✉), Bolier Torres[3,4], Francisco Cabrera-Torres[5,6], Eulalia Vasco[7], Carlos G. H. Díaz-Ambrona[2], and Theofilos Toulkeridis[8,9]

[1] Facultad de Ciencias Agropecuarias, Universidad Técnica Estatal de Quevedo (UTEQ), Quevedo, Ecuador
mageher@gmail.com
[2] AgSystems, Ceigram, itdUPM, Centro de Innovación en Tecnología para el Desarrollo Humano, Universidad Politécnica de Madrid (UPM), 28040 Madrid, Spain
[3] Corporation for Sustainable Development, Conservation and Climate Change, Tena 150150, Ecuador
[4] Departamento de Ciencias de la Vida, Universidad Estatal Amazónica (UEA), Pastaza, Puyo, Ecuador
[5] School of Surveying, Geodesy and Cartography Engineering, Universidad Politécnica de Madrid (UPM), 28040 Madrid, Spain
[6] Instituto Geográfico Militar, Quito 170413, Ecuador
[7] Facultad de Ciencias Agrícolas, Universidad Central del Ecuador, Quito, Ecuador
[8] Universidad de las Fuerzas Armadas ESPE, Sangolquí, Ecuador
[9] Universidad de Especialidades Turísticas UDET, Quito, Ecuador

Abstract. The main aim of the current research has been to quantify and map the pattern of changes in land use coverage (LULC), using free data. The study area has been the northern area of the Diversity and Life Strip, located in the Yasuní Biosphere Reserve, considered to be a biodiversity hotspot. The processes performed in order to apply LULC on the Diversity and Life Zone (DLZ) were 1) data collection and analysis, 2) remote sensing data processing, 3) thematic land cover and 4) transformation from raster to vector. The analysis period was of about 10 years (2009–2018), where Landsat 7–8 satellite images were used. The classified classes were forest, traditional productive system, herbaceous vegetation, infrastructure and water. It was determined that the forest decreased by 3.63%, while the herbaceous vegetation increased by 3.03%. Therefore, the preliminary data from this study provide a benchmark for decision makers and take action in order to minimize deforestation.

Keywords: LULC · Remote sensing · Biodiversity · Yasuni · Ecuador

1 Introduction

Currently there are major environmental problems, such as global warming and the water crisis, which mainly cause adverse impacts on worlds natural resources. In order

S. Berrezueta and K. Abad (Eds.): *Doctoral Symposium on Information and Communication Technologies - DSICT*, LNEE 846, pp. 104–115, 2022.
https://doi.org/10.1007/978-3-030-93718-8_9

to counteract this situation, it is necessary to preserve ecosystems, such as forests, which are fundamental in the collection, conservation, purification and replacement of water [1]. Also, forests absorb and fix CO_2, regulating the global climate. Therefore, as the climate changes, the water resource does too, causing a significant increase in glacial melting, sea level rise, droughts and torrential rains that affect the global ecosystem. Similarly, the oceans alter the modulation of CO_2 levels, varying the global temperature and the impact on transformation of nutrients for marine species. Therefore, the knowledge obtained from scientific research is essential in order to continue with the conservation of ecosystems and improvement of the quality of human life [2].

Although the effects of climate change in Ecuador are widespread [3], there are different perceptions of its impacts at the local level [3–5]. One of the geographical areas, such as the Amazonian lowland, is more vulnerable to climate change, as there is a high pressure on the ecosystems that threatens their stability, due to anthropic activities, mainly due to deforestation. The phenomenon of deforestation causes water deficit, low retention and alteration in humidity that in turn changes the coverage and use of land, encouraging the alteration of radiation [6].

In order to determine the impact of the different anomalies mentioned, there are different satellite platforms such as LANDSAT, or airplanes that conduct the necessary devices to capture, store and transmit a set of free data, agglomerated in satellite images [7], which than provide answers to many questions, providing hereby a point of reference [8]. With such a large amount of information, the data can be shaped to identify problems in a more understandable way [9]. The analysis of free and massive data allows to generate new opportunities for decision makers or researchers, and they have been used in settings of natural importance such as the Galapagos Islands [10], Sumaco Biosphere Reserve [11], Ecuadorian Andes [12], Isla Puna [13], in phenomena of global significance such as SARS-CoV-2 infection among susceptible populations in Ecuador [14], and in crops such as African palm [15]. Based on the background, the main objective of the current research has been to quantify and map the pattern of changes in land use coverage, using free data.

2 Materials and Methods

2.1 Geographical Setting

The study has been realized in the northern area of the Diversity and Life Zone (DLZ), which overlaps in the parishes of Taracoa (787 ha), Dayuma (19,900 ha) of the Francisco de Orellana canton in the province of Orellana, northeastern Ecuador (Fig. 1). The geographical limits to the northern side of the DLS is the Indiyama river, to the south it is the Inés Arango parish, while to the east the limits are the Yasuní National Park (PNY) with some 980,000 ha [17]. The study area belongs to the Yasuní Biosphere Reserve (RBY), covering an area of 2,700,000 ha, which were established as a biosphere reserve by the UNESCO Man and Biosphere program in 1989 [16]. In the RBY, 99.73% is represented by original natural vegetation, where its central conservation area is the PNY [17], being considered a biodiversity hotspot [18]. The RBY overlaps 21 oil fields, extending between the provinces of Orellana (50.51%), Pastaza (39.40%), Napo (8.64%) and Sucumbíos (1.45%),

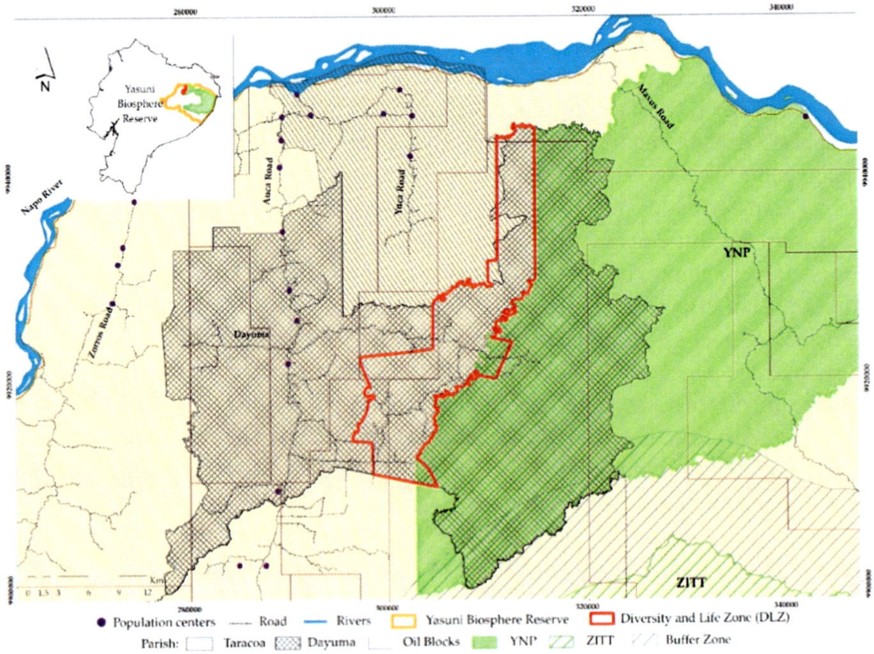

Fig. 1. Location of the diversity and Life Zone in the Yasuni Biosphere Reserve, Eastern Ecuadorian Amazonian lowland.

2.2 Methodology

Next, we will address particularly all the processes applied in the LULC on the DLZ, being 1) data collection and analysis, 2) remote sensing data processing, 3) thematic land cover and 4) transformation from raster to vector (Fig. 3).

Data Collection and Analysis. Landsat satellite images were obtained, which have a high technological development [19], with an altitude of 705 km, inclination of 98° and with a periodicity of 16 days. The LULC spatial dynamics was analyzed for a time interval of approximately 10 years (2009–2018) and have been downloaded from the site of the United States Geological Survey (USGS) based on the Earth Explorer (https://earthexplorer.usgs.gov/), with a spatial resolution of 30 m (Table 1). the selection of the images was performed while trying to avoid the presence of clouds as much as possible, considering that the ecuadorian Amazon Region (EAR) is in the Intertropical convergence zone, which is why it represents the highest cloud cover of the planet [20].

Remote Sensing Data Processing. The images were standardized, below the same coordinate system (SIRGAS, UTM Zone 18S/EPSG: 31993), and corrected, eliminating the presence of anomalies, thus allowing data to be available as closely as possible to an ideal acquisition [21]. The first step has been to conduct a radiometric correction, which allows the passage from digital counts to physical magnitudes, specifically with values of radiances (W m^{-2} sr^{-1} μm^{-1}), and then, using radioactive transfer models, mitigate the atmospheric effect, allowing to obtain surface reflectivity values. Only for the Landsat 7 image, due to the banding it presents, a simple interpolation was applied, which replaces the empty pixels with their neighbors, calling this technique the Neighborhood Similar Pixel Interpolator (NSPI) [22].

Thematic Land Cover. Prior to the extraction of information *per se*, the thematic legend was structured in a scale of 1:100,000, with the classes forest, traditional production system, herbaceous vegetation, infrastructure and water. This has been similar to level II of the legend elaborated by public institutions such as the former Ministry of Agriculture, Livestock, Aquaculture and Fisheries (MAGAP) being currently the Ministry of Agriculture and Livestock (MAG), the Ministry of Environment, Water and Ecological Transition and the former Ecuadorian Space Institute (IEE) being the current Military Geographical Institute (IGM) [23, 24]. Subsequently, the selection of training samples was conducted using as secondary information, the map of use and coverage prepared by SIGTIERRAS (http://ide.sigtierras.gob.ec/geoportal/), on a scale of 1:25,000. In addition, object-based segmentation was applied to the input images, prior to the application of the classification algorithm, and, on the other hand, spectral signatures were obtained (Fig. 2), mitigating to the maximum in this way the error that may occur when selecting the samples.

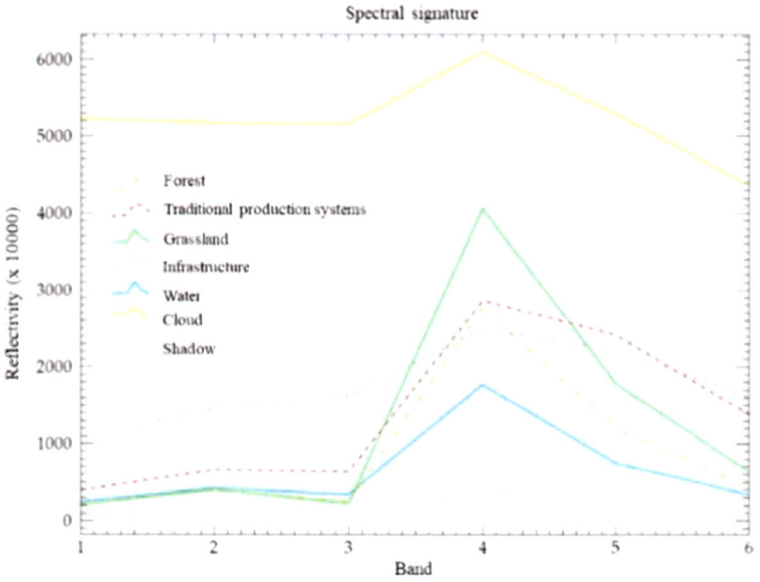

Fig. 2. Typical spectral signature resampled to Landsat - Diversity and Life Zone, Yasuni Biosphere Reserve

The training samples collected are spatially distributed throughout each of the images, and in terms of quantity, it was considered between 10th and 100th pixels [25], with "a" being the number of stripes, therefore, the number of pixels should fluctuate between at least 60 and 600 for each class. These samples were evaluated with the Jeffries Matusita index.

Once the key information had been collected, we proceeded with the supervised classification called neural networks, and through test samples (different from training samples), the precision of the results obtained was evaluated, through the parameters: global precision (OA) and kappa index (K).

Raster to Vector Transformation. Finally, a smoothing filter was applied to the classified images in order to obtain greater homogenization and eliminate possible noise, and the product obtained was converted to vector format. For scales 1: 100,000, the minimum mappable unit is 5 ha, however, to avoid a loss of the information extracted, 1 ha has been taken as the minimum unit [26].

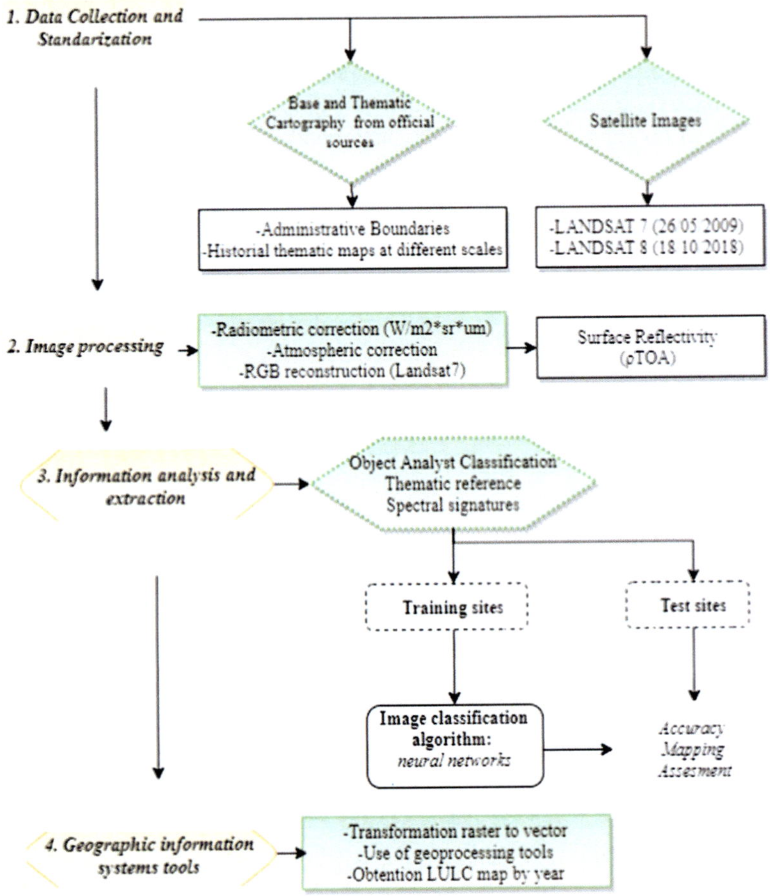

Fig. 3. Workflow for the LULC methodology, in the period 2009–2018

3 Results and Discussion

3.1 Change in LULC in the DLZ 2009–2018

Next, the validation of the LULC classifications and the analysis in the Diversity and Life Zone are detailed (Fig. 1).

Classification Validation. LULC classification levels for dates range from 98.0% to 99.6%, with Kappa concordance rates between 0.97 and 0.99. Accuracies by individual LULC class, that is, User Accuracy (UA) and Producer Accuracy (PA) are listed in Table 2. These Kappa values are satisfactory for the study area as they comply the minimum accuracy of 85%, as stipulated by the Anderson classification scheme [27]. These results provide an important platform for subsequent analysis of LULC changes.

The spectral separability index applied to the training samples, Jeffries Matusita, indicated distances above 1.8 between classes, which means that, when executing the classification algorithm, the confusion is minimal.

Table 1. Satellite images used in spatial analysis LULC

Mission	Sensor	Strip	Path/Row	Date
Landsat 7	ETM+	1, 2, 3, 4, 5, 7	9/60 y 61	26/05/2009
Landsat 8	OLI	2, 3, 4, 5, 6, 7	9/60 y 61	18/10/2018

Table 2. Evaluation of the accuracy of the LULC classification in the DLZ.

Classes	2009		2018	
	PA (%)	UA (%)	PA (%)	UA (%)
Forest	100.00	97.94	100.00	100.00
Traditional productive system	97.40	99.47	100.00	97.78
Herbaceous vegetation	96.02	100.00	100.00	100.00
Infrastructure	96.06	92.42	96.91	100.00
Water	93.85	88.41	100.00	100.00
OA (%)	98.0		99.6	
K	0.97		0.99	

The precision values could be overestimated for the year 2018, however, a comparison was performed with the map of use and coverage of SIGTIERRAS, prepared in 2015, obtaining a degree of similarity of 88.7% (34,472.19 ha of 38,857.5 ha of DLZ). It should be noted that SIGTIERRAS operated with orthophotos and used the visual interpretation methodology.

3.2 Changes in Land Use and Cover (LULC)

In Table 3 and Table 4, together with Fig. 4, summarize the LULC change trend from 2009 to 2018 based on five classes drawn from the North Zone of the DLZ. Furthermore, the spatial representation of the LULC types from 2009 to 2018 is illustrated in Fig. 5. In 2009, the LULC pattern as a percentage of the total area studied was dominated by forest, covering 90.79% of the total area studied, followed by herbaceous vegetation (4.71%), traditional production systems (2.53%), water (1.05%), infrastructure (0.59%).

Trend changes were observed for all LULC in the 2018, where the forest decreased by 3.63%, which is consistent with the dynamics of deforestation that is increasing in tropical forests [28, 29]. Ecuador, Peru and Bolivia belong to the 15 countries with the greatest loss of primary forest area between 1990 and 2015 [30].

In the RAE, the perimeters of protected areas and intangible areas are more vulnerable to deforestation [31] and intensify with oil activity [32, 33], access to road systems [34, 35], and therefore conversion to agriculture [36, 37]. They are also related to the dynamics of wood supply from small producers (mestizo-colonists and indigenous people) to urban centers [38].

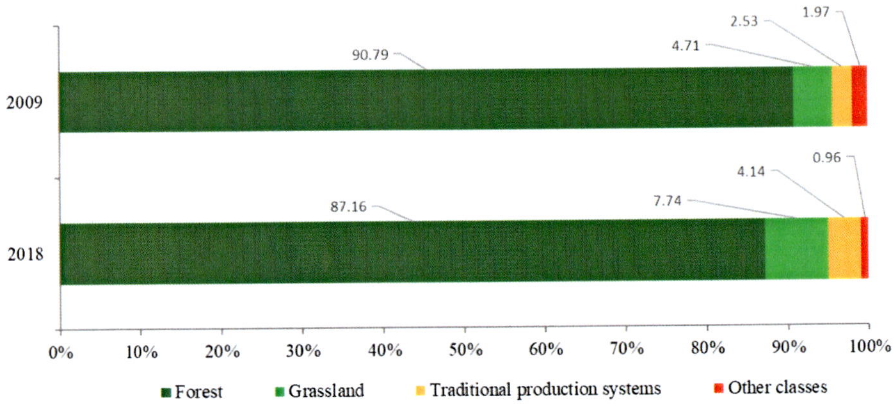

Fig. 4. LULC change graph for 2009, and 2018 at North Zone of the DLZ

In relation to the classification of traditional production systems for 2018, there was an increase of 1.61% (Table 4) and it is concentrated near highways (Fig. 5), which has been driven by colonization, the different state laws and the beginning of oil extraction, promoting migration to the Ecuadorian Amazon and its impact on land use [39], driving deforestation and changes in land use [4]. This LULC dynamic has been the main source of food and sustenance for rural households, a condition very similar to what occurred in the Brazilian Amazon during its colonization processes [41].

The expansion of Grassland type areas in the period studied (Table 3) and their difference (Table 4) indicates an increase of 3.03%, this may reflect the change of stage from subsistence agriculture to agriculture of market in colonial and indigenous populations, driven by diversification of economic and agricultural risk [42] and may also be influenced by the precarious sustainability of traditional production systems kinkore (Waorani), Aja (Shuar), Chakra (Kichwa) and agroforestry systems (migrant settlers) evaluated under the social, environmental, economic and governance dimensions [43–46].

In terms of infrastructure, the results obtained (Table 3) and (Table 4) and represented in Fig. 5, are corroborated with the dynamics analyzed in the Dayuma parish, in the period 1982, 1990, 2000 and 2014 [47], where the expansion of road systems and colonist

Table 3. LULC classification results for 2009 and 2018 images demonstrating each class area and percentages by class at North Zone of the DLZ

Classes	Area 2009		Area 2018	
	ha	%	ha	%
Forest	19052,13	90,79	18290,13	87,16
Traditional productive system	531,48	2,53	868,36	4,14
Grassland	989,21	4,71	1625,22	7,74
Infrastructure	123,39	0,59	72,12	0,34
Water	220,05	1,05	129,49	0,62
Cloud	40,92	0,20	0,00	0,00
Cloud shadow	28,14	0,13	0,00	0,00
TOTAL	20985,33	100,00	20985,33	100,00

Table 4. Results of the LULC classification for 2009 and 2018 images indicating the area changed and percentage at North Zone of the DLZ

Classes	2009–2018 (Area)	
	ha	%
Forest	−762,00	−3,63
Traditional productive system	336,87	1,61
Grassland	636,01	3,03
Infrastructure	−51,27	−0,24
Water	−90,56	−0,43

and indigenous settlements is evidenced, as a result of oil activity and migration and colonization processes [47].

Similar dynamics have occurred in Machadinho D'Oeste, which is a municipality located in the state of Rondônia, in the southwestern part of the Brazilian Amazon. Machadinho is located within the so-called "deforestation arc", being the area with the highest historical rates of conversion of primary forest in the Brazilian Amazon [48]. It has been evidenced that rapid population growth in tropical countries will significantly increase the intensity of the deforestation by 2030 [49]. Therefore, this poses a great challenge for professionals seeking to balance rural development and conservation policies in indigenous territories and areas of mega diversity [50].

The variation of the water class (surface $< 2\%$ of the total DLZ and interannual variation $< 0.5\%$), is mainly due to the complexity at the time of detecting it in the satellite image. This occurs since the spatial resolution of the image is 30 m versus 60 m, which is the average width of the meandering rivers in the study area, requiring several pixels to identify an object.

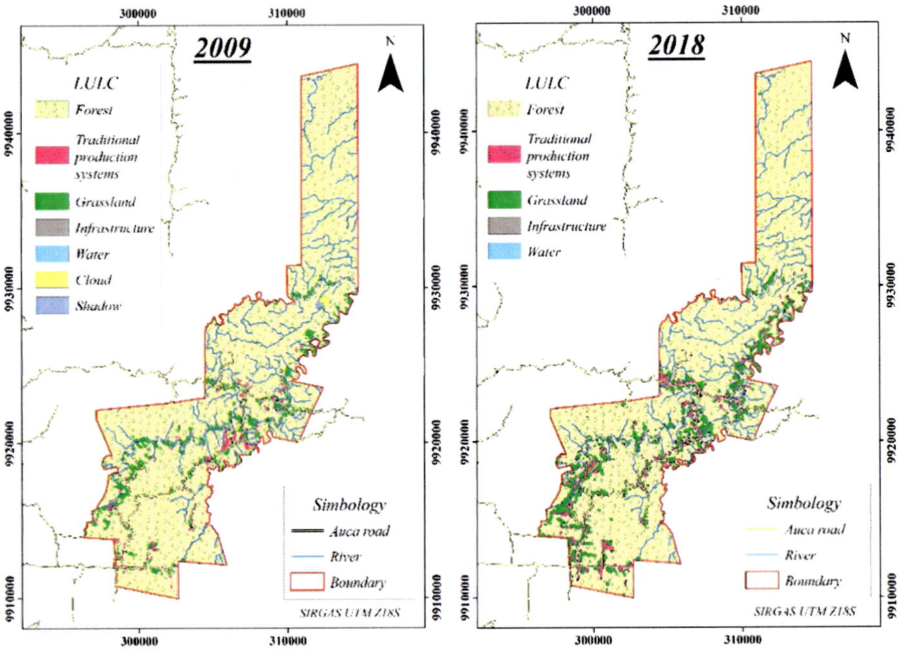

Fig. 5. LULC maps for 2009 and 2018 at North Zone of the DLZ

4 Conclusions

The free data processing with tools such as remote sensing, encourages decision makers to generate medium or large-scale projects, at a minimum cost in order to obtain a reference point for the management of natural resources as public goods and achieve its sustainability.

The low-cost change analysis based on remote sensing images allowed to quantify and map the pattern of change in LULC in the Diversity and Life Zone of the Yasuní Biosphere Reserve and provides a basis for strategic planning, management and planning. decision-making for the conservation of natural resources, improve the life strategies of indigenous communities and migrant settlers and can be a tool to safeguard the advance of the agricultural frontier within the Yasuní National Park.

References

1. Montiel-González, I., Martínez-Santiago, S., López, S.A., García, H.G.: Impacto del cambio climático en la agricultura de secano de Aguascalientes, México, para un futuro cercano (2015–2039). Revista Chapingo Serie Zonas Áridas **16**(1), 1–13 (2017)
2. IPCC. Resumen para responsables de políticas. En: El cambio climático y la tierra: Informe especial del IPCC sobre el cambio climático, la desertificación, la degradación de las tierras, la gestión sostenible de las tierras, la seguridad alimentaria y los flujos de gases de efecto invernadero en los ecosistemas terrestres (2019)
3. Toulkeridis, T., et al.: Climate Change according to Ecuadorian academics–Perceptions versus facts. LA GRANJA. Revista de Ciencias de la Vida **31**(1), 21–46 (2020)

4. León Alvear, V, et al.: Percepción sobre cambio climático en cuatro comunidades orientadas a la ganadería bovina en la zona central de los Andes Ecuatorianos. Livestock Research for Rural Development, vol. 32, Article #165 (2020)
5. Heredia, M., Barreto, D., Toulkeridis, T.: Percepción de las poblaciones indígenas kichwa sobre el cambio climático y sus implicaciones en Puyo, Ecuador. En La acción climática en las ciudades latinoamericanas: aproximaciones y propuestas Publisher: FLACSO Quito, Ecuador (2020)
6. Tiria, F.L.N., Bonilla, C.J.S., Bonilla, C.C.A.: Transformación de las coberturas vegetales y uso del suelo en la llanura amazónica colombiana: el caso de Puerto Leguízamo, Putumayo (Colombia). Cuadernos de Geografía: Revista Colombiana de Geografía **27**(2), 286–300 (2018)
7. Masek, J.G., et al.: Landsat 9: empowering open science and applications through continuity. Remote Sens. Environ. **248**, 111968 (2020)
8. Rocchio, L.E., Davis, A.L.: Bridging the divide: translating landsat research into usable science. In: AGU Fall Meeting Abstracts, vol. 2006, pp. PA31A–0827 (2006)
9. Craglia, M., Ostermann, F., Spinsanti, L.: Digital earth from vision to practice making sense of citizen-generated content. Int. J. Digit. Earth **5**(5), 398–416 (2012)
10. Barreto-Álvarez, D.E., Heredia-Rengifo, M.G., Padilla-Almeida, O., Toulkeridis, T.: Multi-temporal evaluation of the recent land use change in Santa Cruz Island, Galapagos, Ecuador. In: Rodriguez Morales, G., Fonseca C., E.R., Salgado, J.P., Pérez-Gosende, P., Orellana Cordero, M., Berrezueta, S. (eds.) TICEC 2020. CCIS, vol. 1307, pp. 519–534. Springer, Cham (2020). https://doi.org/10.1007/978-3-030-62833-8_38
11. Torres, B., Andrade, L., Navarrete, A.T., Vasco, C., Robles, M.: Cambio de uso del suelo en paisajes agrícolas-forestales: análisis espacial en cinco comunidades Kichwas de la Región Amazónica Ecuatoriana. Revista Amazónica Ciencia y Tecnología **7**(2), 105–118 (2018)
12. Guascal, E., Rojas, S., Kirby, E., Toulkeridis, T., Fuertes, W., Heredia, M.: Application of remote sensing techniques in the estimation of forest biomass of a recreation area by UAV and RADAR images in Ecuador. In: 2020 Seventh International Conference on eDemocracy & eGovernment (ICEDEG), pp. 183–190. IEEE (2020)
13. Pérez, K.A., Lascano, P.P., Sánchez, I.M., Padilla-Almeida, O., Toulkeridis, T.: Evaluation of the surface temperature applied in aquaculture based on satellite images in Coastal Ecuador. In: Rodriguez Morales, G., Fonseca C., E.R., Salgado, J.P., Pérez-Gosende, P., Orellana Cordero, M., Berrezueta, S. (eds.) TICEC 2020. CCIS, vol. 1307, pp. 572–586. Springer, Cham (2020). https://doi.org/10.1007/978-3-030-62833-8_42
14. Toulkeridis, T., Seqqat, R., Arias, M.T., Salazar-Martinez, R., et al.: Volcanic ash as a precursor for SARS-CoV-2 infection among susceptible populations in Ecuador: a satellite Imaging and excess mortality-based analysis. Disaster Med. Public Health Prep. 1–37. (2021)
15. Viera-Torres, M., Sinde-González, I., Gil-Docampo, M., Bravo-Yandún, V., Toulkeridis, T.: Generating the baseline in the early detection of bud rot and red ring disease in oil palms by geospatial technologies. Remote Sens. **12**(19), 3229 (2020)
16. UNESCO. Yasuni Biosphere Reserve: Biosphere reserves in Latin America and the Caribbean. https://en.unesco.org/biosphere/lac/yasuni. Accessed 15 June 2021
17. Taco, M.P.: El Parque Nacional Yasuní. In: Jorgenson JP, Rodríguez MC, editors. Conservación y desarrollo sostenible del Parque Nacional Yasuní y su área de influencia. Quito, Ecuador: Ministerio del Ambiente/UNESCO/WCS, pp. 48–51 (2001)
18. Finer, M., Vijay, V., Ponce, F., Jenkins, C. N., Kahn, T.R.: Ecuador's Yasuni Biosphere Reserve: a brief modern history and conservation challenges. Environ. Res. Lett. **4**(3), 034005 (2009)
19. Sobrino, J., et al.: Teledetección. 467 pp. Servicio de Publicaciones Universidad de Valencia. Valencia, España (2000)

20. Vargas, G.: La tropicalidad y el análisis geográfico. Reflexiones, vol. 81. Costa Rica. (2002)
21. Chuvieco, E.: Teledetección ambiental, 591 pp. Editorial Planeta. España, Barcelona (2010)
22. Saldarriaga, L.: Reconstrucción de base de datos Landsat7 ETM+ SLC-off para cuantificar, detectar cambios y cartografiar la cobertura vegetal del valle bajo del río Chira – Perú, durante el período 2005–2014. 68 pp. TFM, Máster en Teledetección, Universidad de Valencia. Valencia, España (2014)
23. MAE. Línea base de deforestación del Ecuador Continental, 32 pp. Disponible en. http://sociobosque.ambiente.gob.ec (2012)
24. IPCC. Guidelines for National Greenhouse Gas Inventories - Volume 4: Agriculture, Land Use and Forestry (AFOLU). http://www.ipcc-nggip.iges.or.jp (2006)
25. Jensen, J.: Introductory Digital Image Processing: A Remote Sensing Perspective. 316 pp. Editorial Prentice Hall. Michigan, Estados Unidos (1996)
26. Lencinas, J., Siebert, A.: Relevamiento de bosques con información satelital: Resolución espacial y escala. Quebracho **17**, 101–105 (2009)
27. Anderson, J.R.: A Land Use and Land Cover Classification System for Use with Remote Sensor Data; Geological Survey Professional Paper 964; US Government Printing Office: Washington. DC, USA (1976)
28. Sloan, S., Sayer, J.A.: Forest resources assessment of 2015 shows positive global trends but forest loss and degradation persist in poor tropical countries. For. Ecol. Manage. **352**, 134–145 (2015)
29. FAO, F.: Global Forest Resources Assessment 2015. How Are the World's Forests Changing? FAO, Rome (2015)
30. Morales-Hidalgo, D., Oswalt, S.N., Somanathan, E.: Status and trends in global primary forest, protected areas, and areas designated for conservation of biodiversity from the Global Forest Resources Assessment 2015. For. Ecol. Manage. **352**, 68–77 (2015)
31. Heredia-R., M., et al.: Towards the Sustainability of Traditional Agroforestry Systems Kichwa. Sumaco Biosphere Reserve Case, Amazonia (2021). https://doi.org/10.3390/IECAG2021-09733
32. Llerena-Montoya, S., Velastegui-Montoya, A., Zhirzhan-Azanza, B., et al.: Multitemporal analysis of land use and land cover within an oil block in the Ecuadorian Amazon. ISPRS Int. J. Geo Inf. **10**(3), 191 (2021)
33. Mena, C.F., Laso, F., Martinez, P., Sampedro, C.: Modeling road building, deforestation and carbon emissions due deforestation in the Ecuadorian Amazon: the potential impact of oil frontier growth. J. Land Use Sci. **12**(6), 477–492 (2017)
34. Davidson, E., de Araújo, A., Artaxo, P., et al.: The Amazon basin in transition. Nature **481**, 321–328 (2012). https://doi.org/10.1038/nature10717
35. Rorato, A.C., Picoli, M.C., Verstegen, J.A., Camara, G., Silva Bezerra, F.G.S, Escada, M.I.: Environmental Threats over Amazonian Indigenous Lands. Land **10**(3), 267 (2021)
36. Viteri-Salazar, O., Toledo, L.: The expansion of the agricultural frontier in the northern Amazon region of Ecuador, 2000–2011: process, causes, and impact. Land Use Policy **99**, 104986 (2020)
37. Junquera, V., Meyfroidt, P., Sun, Z., Latthachack, P., Grêt-Regamey, A.: From global drivers to local land-use change: understanding the northern Laos rubber boom. Environ. Sci. Policy **109**, 103–115 (2020)
38. Mejia, E., Pacheco, P., Muzo, A., Torres, B.: Smallholders and timber extraction in the Ecuadorian Amazon: amidst market opportunities and regulatory constraints. Int. For. Rev. **16**, 1–13 (2015)
39. Maldonado-Lince, G.L.: reforma agraria en el Ecuador, una lucha por la justicia. Nueva sociedad **41**, 14–29 (1979)

40. Viteri-Salazar: O.: Evaluación de la sostenibilidad de los cultivos de café y cacao en las provincias de orellana y sucumbios-ecuador. Doctoral dissertation. Universitat Autònoma de Barcelona (2013)
41. Walker, R.: Mapping process to pattern in the landscape change of the Amazonian frontier. Ann. Assoc. Am. Geogr. **93**(2), 376–398 (2003)
42. Carr, D.L., Suter, L., Barbieri, A.: Population dynamics and tropical deforestation: state of the debate and conceptual challenges. Popul. Environ. **27**(1), 89–113 (2005)
43. Heredia, M., Bravo, C., Torres, B., Alemán, R.: Innovación para el fortalecimiento de capacidades sobre sostenibilidad de los recursos naturales en poblaciones indígenas y mestizas–colonas: Reserva de Biosfera Yasuní. Revista Ibérica de Sistemas e Tecnologias de Informação, (E25), pp. 103–116 (2005)
44. Heredia, M., Torres, B., Alemán, R., Bravo, C., Hernández Díaz-Ambrona, C.: SAFA and GeoGebra allies to evaluate natural and cultural sustainability: Yasuni biosphere reserve. Int. J. Adv. Sci. Eng. Inf. Technol. **10**(2), 880–887 (2020)
45. Heredia, M., Falconí, A.K., Barreto, D., Amores, K., Jamil, H., Torres, B.: Conductas sustentables sobre el marco de evaluación SAFA-FAO: un aporte para poblaciones rurales vulnerables de la Amazonía. Revista Ibérica de Sistemas e Tecnologias de Informação **E33**, 312–326 (2020)
46. Heredia-R, M., et al.: Evaluación de la sostenibilidad de sistemas productivos en la franja de diversidad y vida: reserva de biosfera yasuní, amazonía. Revista Científica Axioma **22**, 5–11 (2020)
47. Houssou, N.L.J., Cordero, J.D., Bouadjio-Boulic, A., et al.: Synchronizing histories of exposure and demography: the construction of an agent-based model of the Ecuadorian Amazon colonization and exposure to oil pollution hazards. J. Artif. Soc. Soc. Simul. **22**(2). (2019)
48. Barbieri, A.F., Guedes, G.R., Dos Santos, R.O.: Land use systems and livelihoods in demographically heterogeneous frontier stages in the amazon. Environ. Dev. **38**, 100587 (2021)
49. Vieilledent, G., Grinand, C., Vaudry, R.: Forecasting deforestation and carbon emissions in tropical developing countries facing demographic expansion: a case study in Madagascar. Ecol. Evol. **3**(6), 1702–1716 (2013)
50. Vasco, C., Bilsborrow, R., Torres, B., Griess, V.: Agricultural land use among mestizo colonist and indigenous populations: contrasting patterns in the Amazon. PLoS ONE **13**(7), e0199518 (2018). https://doi.org/10.1371/journal.pone.0199518

Data and Text Mining for the Detection of Fraud in Public Contracts: A Case Study of Ecuador's Official Public Procurement System

Yeferson Torres-Berru[1,2(✉)] and Vivian Felix López Batista[1]

[1] Universidad de Salamanca Plaza de la Merced, Salamanca, Spain
{ymtorresb,vivian}@usal.es
[2] Instituto Superior Tecnológico Sudamericano, Loja, Ecuador

Abstract. Corruption is present in different forms and typologies, directly affecting the execution of both public and private contracts. The doctoral thesis aims to establish a methodology to prevent and detect corruption automatically in public procurement. By using machine learning techniques and Natural Language Processing (NLP), algorithms for detecting and predicting favouritism and oligopoly are developed. In addition to detecting corruption and its types in the Ecuadorian Public Procurement System (SERCOP) and also visualising the results in an appropriate way, in order to detect and prevent future acts of corruption. In order to analyse the feasibility of the study, a mapping and systematic literature review was carried out, allowing the hypothesis and the methodology to be followed in order to execute and evaluate the developed algorithms. Finally, the detection of favouritism based on process qualification parameters and types of contracting is tested.

Keywords: Corruption · Public procurement · Data mining · Machine learning

1 Introduction

Transparency International estimates that the costs of corruption in public procurement amount to 20–25% of the value of the contract, and can sometimes reach 40–50% [1]. According to the Inter-American Development Bank (IDB), on average, public procurement accounted for 32.5% of general government spending, 29.8% in Latin American countries and 8.6% of gross domestic product (GDP) in the Caribbean. However, the size of expenditure on this item varies roughly from 15 to 47%, due to the higher share of capital expenditure in total expenditure. Ecuador is the third country in the region in capital expenditure,

Doctorado en Ingeniería Informática Universidad de Salamanca.

S. Berrezueta and K. Abad (Eds.): *Doctoral Symposium on Information and Communication Technologies - DSICT*, LNEE 846, pp. 116–127, 2022.
https://doi.org/10.1007/978-3-030-93718-8_10

with a figure close to 45%, as well as being the first country in the region in terms of public procurement in terms of GDP, with 16% [2,3].

In Ecuador, SERCOP is in charge of promoting citizen participation, increasing access to and use of public information by the population. Increasing transparency, combating fraud and corruption that could arise from malpractices in public procurement. The platform provided by SERCOP contains documents in PDF format for each contracting process, where Data on the Specifications (TDR in Spanish), invitations to suppliers, bids made, observations, in short, all the documentation generated by the purchase are stored. Among the processes present in SERCOP, the following are available:

- Execution of works.
- Procurement of products and services.
- Contracting of the consultancy.

Different forms and levels of corruption have emerged, namely bribery, embezzlement, fraud, extortion, breach of trust, collusion and favouritism [5,8]. The main corruption mechanisms according to [6,7] are: non-existence of contract, inappropriate contracting, fractioning, contract modifications. As can be seen, the study of corruption covers a wide range of social and human sciences, with theoretical and scientific contributions.

It is also established that for there to be corruption in a process, the following factors must be taken into account [9,10]:

- The type of product, the amount of contracting and the type of purchase.
- The bidding period and validity.
- Modifications during the execution of the process (sheets, parameters, questions and answers).
- Changes in the percentages in the qualifying parameters.
- The personal relationships between persons of the contracting companies.
- The specific experience required of a supplier.
- The detailed technical specifications.

The data mining plays an important role in corruption detection, and is most commonly used to find hidden information in large amounts of data [11]. The corruption case known as *Panama papers* [12] revealed tax and financial fraud to the public opinion; in Brazil the Observatory of Public Expenditures [13] reviewed more than 120,000 public contracts and uncovered more than 7,500 cases involving \$ 104 million in financial operations of dubious legality. These examples illustrate the importance of data science in the fight against corruption. This doctoral thesis aims to implement a methodology to prevent and detect corruption automatically, in public procurement in SERCOP, with the use of Artificial Intelligence (AI) techniques such as machine learning and PLN.

2 Research Work Development

In order to form the state of the art on the investigation topic, we conducted a systematic bibliography search. We followed the methodology proposed by [14], that divides the process into three phases: planning, carrying out the review and elaboration of the paper. As a product of this, two articles were written and indexed in SCOPUS (Quartile 3).

Work 1

Systematic mapping was developed [15] of scientific publications (2015–2019), centered on contractual corruption in its various forms, by applying data mining and machine learning techniques. We present six research questions to answer the analysis of 147 articles obtained from the Web of Science (WoS) and Scopus databases. The detection of fraud, financial fraud and corruption predominate in the investigations, the most common forms of corruption are fraud (72.72%) and overpricing (8.84%). The investigations were carried out in the United States (16.32%), China (10.88%), the United Kingdom (8.94%) and in Latin America, mainly in Brazil (3.4%), with minimal contributions from Colombia and Paraguay. Figure 1 shows a survey of the articles consulted according to the country in which they were published and the type of corruption.

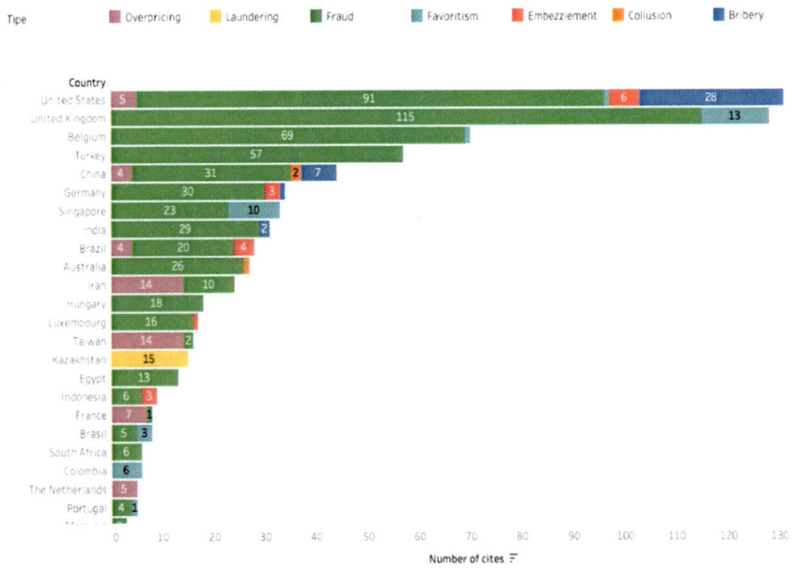

Fig. 1. Articles published by countries and type of corruption.

Work 2

In their study Torres-Berru et al. [16] analyzes different types of corruption (bribery, collusion, embezzlement, fraud, abuse of discretionary power, favoritism, nepotism) and six types of IA techniques (classification, regression, *clustering*, prediction, outlier detection and visualization). The methodology proposed by Torres-Carrión was used, [14], and four investigation questions were posed to find out the types of searches performed, the characteristics of the organizations in which the investigations are carried out, the technological tools and the data mining methodologies and techniques. The review was conducted in the Scopus and WoS databases, obtaining 108 articles published between 2015 and 2019.

As a result of this first phase of documentation, as part of the development of the research work, we can summarize that the *Web Scraping* It is a little-used technique to obtain data on corruption studies in contracts. Its use can serve as a basis for future data collection. The few works related to contract analysis in public procurement use isolated data sets and do not consider documents as an initial basis for analysis. It is also evident that the software tools developed for the analysis of corruption in contracts, both in the public and private sectors, are not considered as computer security standards, and the percentage of tools in the web environment is very low.

The main techniques of AI found are logistic models, neural networks, Bayesian networks and support vector machine. The Fraud Score is proposed as a specific metric for assessing corruption risk. Also taken into account are the metrics used to evaluate classification within machine learning, based on the matrix of confusion and (*Receiver Operating Characteristic*) the ROC curves. In addition, supervised learning is the most widely used technique when applying machine learning models in this area.

2.1 Thesis Objectives

Taking into account the theoretical basis discussed in the previous section, it is chosen to work on favouritism, for in [15] work sit is evident that out of 147 articles only 8 deal with this type of corruption.

The objective is focused on establishing a methodology to prevent and detect corruption automatically, through the use of algorithms for the detection of corruption in public procurement. Using machine learning techniques and PLN, the aim is to develop algorithms for detecting and predicting favouritism and oligopoly in public procurement.

In addition, the following **hypothesis** was formulated.

The evaluation of the data and text generated in the phases of a public procurement process facilitates the detection of the presence of corruption, its type, the phase in which it occurs and the detriment to the state.

3 Methodology for Intelligent Discovery of Corruption

After reviewing the different approaches that exist in the current literature on corruption, which attempt to provide an answer to the problem posed, this section details the proposal of the present work, designed to test the hypothesis of the present study. In general, we describe a data mining system that uses PLN to intelligently perform content analysis of contracts and automatically detect corruption. In particular, the two approaches that have been developed and how the experimentation phase is planned to respond to different shortcomings detected in the field of study.

The work aims to assess favouritism and oligopoly as a common form of procurement, which leads to other forms of corruption such as price increases, irregular processes, bribery, etc. Favouritism is the natural human propensity to favour friends, family and anyone close and reliable within a public process, nepotism means the granting of favours to persons who are related to the official holding a public office [8]. To evaluate the favouritism, the variables listed in Table 1 are considered.

Table 1. Variables for assessing favouritism

Variable	Description
Economic offer	Awarding low scores for the evaluation of economic offers, causing high priced purchases to the detriment of the state
Other parameters	Conditions that only one agreed bidder will be able to meet, and to which they will award very high scores
Time	Very brief terms are stipulated for the design, preparation, drafting and submission of proposals. If all the requirements are fulfilled in anticipation, someone will submit the proposal within the established terms
Experience	Entities in order to direct the procedures to the bidders with whom they have previously cooperated before request specific experience with the same entity, which leaves no possibility for new bidders to be awarded
Technical specifications	Institutions include requirements and/or equipment that only satisfy The supplier with which it reached an agreement
Change conditions	The Entity modifies certain parameters in the "Questions, Answers and Clarifications" stage, in order to ensure that the selected supplier's offer obtains the highest scores by including requirements and/or equipment that only that supplier fulfills
Relations between individuals	A group of companies are involved in all tenders of an entity, agreeing beforehand, between them, who is to win the process. In addition, the entity and suppliers arrive at an agreement to generate a kind of oligopoly, where they seek to generate specifications for the benefit of this group of bidders

3.1 Machine Learning Algorithms

The literature review found that most of the published work uses supervised learning, as contracts with anomalies are correctly labelled, 79% of the research corresponds to detection and 21% to prediction. Ecuador's public procurement system does not have an anomalous procurement section. Therefore, lacking labelled data, in the initial phase of the research, the decision was made to use unsupervised learning techniques to detect anomalous patterns in contracts. The proposed methodology is summarised in Fig. 2. Once the retrieved public procurement data is processed, it is analysed through a multi-stage model, which uses different algorithms, like *clustering* (K-Means), Self-organizing map (SOM), Support Vector Machine (SVM) y Deep Learning.

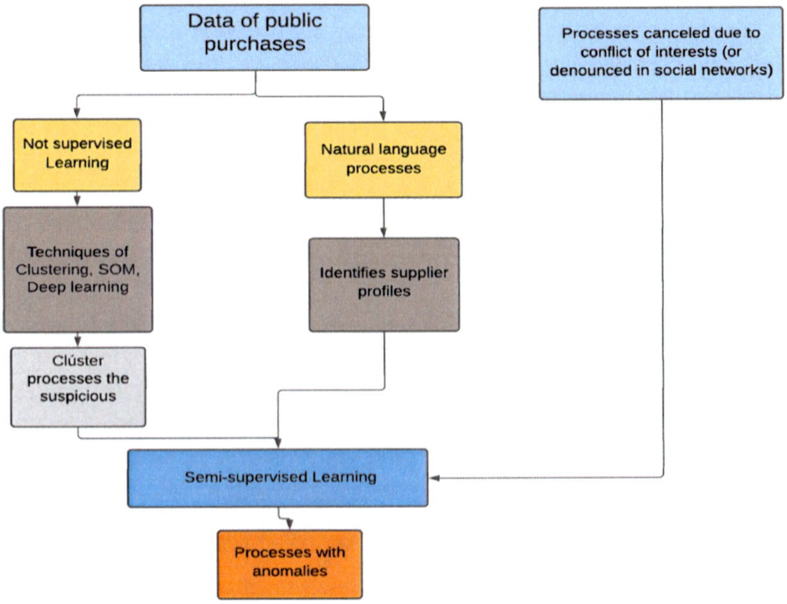

Fig. 2. Anomaly detection model

4 Experimentation

4.1 Data Collecting

In accordance with the proposed methodology, the first step is to retrieve data on public procurement. Because there is no open access portal to the data, the web-based *scraping* technique is applied [19] on the data provided on the website of the National Public Procurement Service of Ecuador[1]. Through this process, information is obtained regarding public processes from 2010 to 2020, as well as the documents (attachments) of each process. Information was retrieved for a total of 1276867 procurement processes.

Considering the process *URL* as input, the different fields of each process are: its description, dates, products, qualification parameters, invitations, files and supplier questions. Each section was extracted according to its equivalent html tag through *scraping* and stored in a non-relational database (MongoDB). Once the data is obtained, it is sorted to remove noise and inconsistency and reduce dimensionality.

[1] https://www.compraspublicas.gob.ec/ProcesoContratacion/compras/.

4.2 Training Phase

Clustering is one of the most popular unsupervised learning techniques. It is used to analyse data and find groups within that data using some kind of similarity measure, such as Euclidean distance. For this we should assume that the number of clusters is known in advance, so we start by setting the cluster number to 8 to be possible to classify our data accordingly. There is no universal similarity metric that works for all cases (it depends on the problem itself). So we iterated to update the centroids until they stopped changing and have been placed in optimal locations to cluster the observations into 8 different centroids. The *elbow* method in $k = 4$ suggests that it is the optimal value for the number of clusters with 10000 iterations to obtain the best result in the evaluation metric.

Self-organising maps (SOM) are unsupervised artificial neural networks based on the winner-takes-all principle [22]. A typical SOM consists of a layer of input neurons, an array of nodes as an output map and an array of connections between each unit of the output layer and all units of the input layer. Input nodes are propagated to a set of output nodes, which are organised into topographic maps, which determine how the spatial location of an output node on the topographic map corresponds to a particular feature of the input data pattern [20]. A rectangular topology consisting of 20 rows and 20 columns is used, allowing individual features to connect to a node and set weights. Each input neuron i is connected to each of the output neurons j by a weight W_{ji}. In this way, the output neurons have an associated vector of weights W_j called the reference vector, allowing the map to make a projection from a multidimensional data space to a two-dimensional map of neurons. To find the proximity between the data, the neighbourhood is evaluated *gaussian* which causes the change of values to decrease with distance and a bubble neighbourhood, which changes all vectors belonging to the neighbourhood to the same shape.

With the 4 clusters obtained (representing the types of contract), semi-supervised learning is applied for anomaly detection (Fig. 3) which is detailed below: the first step is to separate the datasets into a training set (80%) and a test set (20%). For this technique, the processes associated with the clusters where the economic factor is the main factor for the qualification are defined as "normal", subsequently two models are compared:

1. **One-Class Support Vector Machine**. SVM [23] is a supervised type classifier, which is defined by a hyperplane between classes. Given labelled training data and a binary classification problem, the SVM finds the optimal hyperplane that separates the training data into two classes. The algorithm requires training data with two labels, belonging to one of the two classes. The problem appears when you want to apply the algorithm on data where there is a lot of information for one class, but not for the other. In these cases, SVM can be used as an anomaly detector, as a classifier of a class. The model is trained on the data of the class that is considered normal, and data that is different can be predicted. When this version of the algorithm is applied, we have used the property *nu* which allows to control the balance between outliers and normal cases, and therefore assigned *nu*= [1e−3, 1e−2, 1e−1, 1], while the parameter

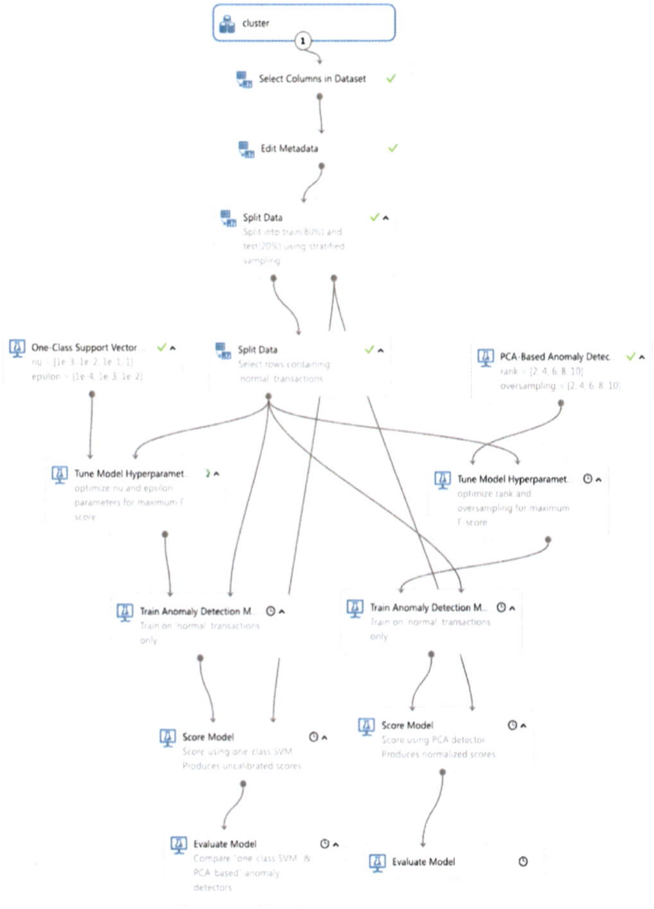

Fig. 3. Azure anomaly detection model

affecting the number of iterations used, when optimising the model, is defined as *epsilon* = [1e−4, 1e−3, 1e−2]. The optimal hyperplanes for machine learning are then determined using a *Hyperparameters* Model. Finally, the model is trained and evaluated using the ROC and *accuracy* metrics.

2. **Principal Component Analysis (PCA)**. The anomaly detection module based on PCA analyses the available features to determine what constitutes a "normal" class, and applying distance metrics to identify the cases that represent anomalies, therefore used with a range of parameters (*rank*) y *oversampling* de [2, 4, 6, 8, 10]. Finally, the model is trained and evaluated using the Score Model, the ROC and *accuracy* metrics.

4.3 Study Cases

In order to validate the functioning of the proposed method, two case studies have been carried out: one to detect anomalies in the qualification parameters in contracts (in the process of publication), allowing the identification of processes in which favouritism exists, and another to investigate price speculation in medical products generated by COVID-19 in Ecuador (published).

Anomaly Detection in Contract Qualification Parameters. To create the dataset for the first case study, the bid scoring parameters were evaluated in 275,730 public procurement contracts in Ecuador, between 2010 and 2020. Twenty-three variables were evaluated to determine the winner of each contract process, among them: economic offer, experience, equipment and instruments, national production, "Other parameters", etc. In this way, it is possible to determine in which processes low scores are awarded to evaluate the economic offer, causing prejudice to the government.

As a result of the experimentation phase, applying SOM and Kmeans, we can highlight that in 3 of the 4 clusters the economic offer is respected as an outstanding qualification parameter.

Note that the semi-supervised learning model applying SVM and PCA can be applied in the evaluation of the regression model and for the detection of anomalies in the processes, taking into account that they mostly belong to the *cluster 4*. As metrics for evaluating the success of the applied algorithms, the following were used: precision (0.95%) and *accuracy* (0.85%).

Exceptional Prices of Medical Supplies During the COVID-19 Pandemic in Ecuador. An exploratory data analysis explores the prices of procurement of supplies, through public procurement contracts in Ecuador, for use in clinical settings or as protection for the general population in response to the COVID-19 pandemic. The study [24] quantifies the differences in the prices of commonly procured medicines and commodities in public procurement contracts identified as related to the COVID-19 pandemic. Statistical analysis was performed to extract relevant measures for each product and to determine variability over all products.

As a result, Ecuador was found to have spent $257 million on public procurement of basic supplies related to COVID-19. Among the most purchased products were masks, paracetamol and PCR tests. Prices varied widely, depending on the individual contract and the number of units purchased. Some prices were exceptionally higher than their market value and as much as 1300% difference with similar purchases. Compared with 2019, the mean price of medical examination gloves increased up to 1,307%, acetaminophen 500 mg pills, up to 796%, and oxygen flasks, 30.8%.

Figure 4 shows the significant price increase in procurement of medical products in April 2020, despite not being as high in demand as in previous months.

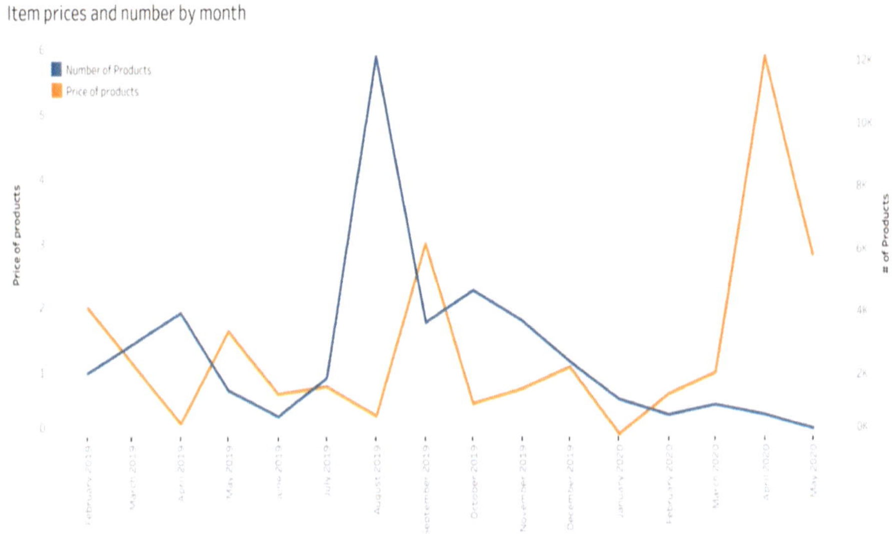

Fig. 4. COVID-19 pandemic medicine price developments.

5 Conclusions

In this article we have summarised the methodology we are developing to prevent and detect corruption in public procurement, using machine and deep learning algorithms. The theoretical basis of the doctoral work has been established, which has led to the publication of three scientific articles. In addition, the few exploitation of favouritism as a form of corruption in current research has been noted. A scientific methodology has been defined according to the hypothesis, based on a hybrid model that includes different phases with supervised and unsupervised learning, PLN and neural networks. This methodology is being tested in two case studies. One related to contract qualification and the other to investigate price speculation in medical products generated by COVID-19 in Ecuador.

6 Future Work

As a future line of work, the aim is to build a *framework* that evaluates, detects and helps in the prediction of favouritism in public procurement processes. In addition to incorporating *Deep Learning* algorithms in the methodology, as well as the NLP for the classification of contractors and relations with the entities, assessing award times.

References

1. Náez Gómez, J.E.: Relación entre el Índice de Control de la Corrupción y algunas variables sociales, económicas e institucionales. Nómadas. Rev. Crítica Ciencias Soc. y Jurídicas 38 (2013)

2. Brito-Gaona, L.F., Iglesias, E.M.: Inversión privada, gasto público, presión tributaria en América Latina. Estududios de Economia. **44**, 5–30 (2017)
3. Izquierdo, A., Pessino, C., Vuletin, G.: Mejor gasto para mejores vidas: Cómo América Latina y el Caribe puede hacer más con menos, vol. 10. Inter-American Development Bank (2018)
4. Servicio Nacional de contratacion publica: Rendicion de cuentas (2018)
5. Moran, J.: Democratic transitions and forms of corruption. Crime, Law Soc. Chang. **36**, 379–393 (2001).https://doi.org/10.1023/A:1012072301648
6. Castro Cuenca, C.G.: La corrupción pública y privada: causas, efectos y mecanismos para combatirla - Google Play (2017)
7. Cassagne, J.C., Rivero Ysern, E.: La contratación pública, Hammurabi (2007)
8. Vargas-Hernández, J.G.: The Multiple Faces of Corruption: Typology, Forms and Levels. SSRN Electron. J. (2009). https://doi.org/10.2139/ssrn.1413976
9. Ponce, H.G., Gil, M.T.N., Durán, M.P.: Responsible public procurement. Des. meas. indicators. CIRIEC-Espana Rev. Econ. Publica, Soc. y Coop. **44**, 253–280 (2019)
10. Subdirección General de Control Coordinación Técnica de Controversias: Manual De Buenas Prácticas En La Contratación Pública Para El Desarrollo Del Ecuador. 1-46 (2015)
11. Alvarez-Jareño, J.A., Badal-Valero, E., Pavia, J.M.: Aplicación de métodos estadísticos, económicos y de aprendizaje automático para la detección de la corrupción. (2019)
12. Woodie, A.: Inside the Panama Papers: How Cloud Analytics Made It All Possible. https://www.datanami.com/2016/04/07/inside-panama-papers-cloud-analytics-made-possible/. Accessed 12 Aug 2019
13. Controladoria-Geral da União: Observatório da Despesa Pública - Controladoria-Geral da União. http://www.cgu.gov.br/assuntos/informacoes-estrategicas/observatorio-da-despesa-publica. Accessed 12 Aug 2019
14. Torres-Carrión, P.V., Gonzalez-Gonzalez, C.S., Aciar,S., Rodriguez-Morales, G.: Methodology for systematic literature review applied to engineering and education. In: IEEE Global Engineering Education Conference, EDUCON 2018-April, pp. 1364–73 (2018)
15. Torres-Berru, Y., López-Batista, V.F., Torres-Carrión, P.: Data mining to detect and prevent corruption in contracts: Systematic mapping review. RISTI - Rev. Iber. Sist. e Tecnol. Inf. **2020**, 13–26 (2020)
16. Torres Berru, Y., López Batista, V.F., Torres-Carrión, P., Jimenez, M.G. : Artificial Intelligence techniques to detect and prevent corruption in procurement: a systematic literature review. In: Botto-Tobar M., et al. (eds) Applied Technologies. ICAT 2019. Communications in Computer and Information Science, vol. 1194. Springer, Cham (2020). https://doi.org/10.1007/978-3-030-42520
17. Hotelling, H.: A generalized T test and measure of multivariate dispersion. In: Proceedings of the Second Berkeley Symposium on Mathematical Statistics and Probability, pp. 23–41 (1951)
18. Ultsch, A., Mörchen, F.: ESOM-Maps: tools for clustering, visualization, and classification with Emergent SOM, pp. 1–7. Tech. Rep. Dept. Math. Comput. Sci. Univ. Marburg Ger (2005)
19. Saurkar, A.V., Gode, S.A.: An overview on web scraping techniques and tools. Int. J. Futur. Revolut. Comput. Sci. Commun. Eng. **4**, 363–367 (2018)
20. Merkl, D.: Text classification with self-organizing maps: some lessons learned. Neurocomputing **211–3**, 61–77 (1998)

21. Vettigli, G.: MiniSom: minimalistic and NumPy-based implementation of the Self Organizing Map
22. Kohonen, T.: Self-organizing Maps. Springer-Verlag, Berlin (1995)
23. Kotsiantis, S.B.: Supervised machine learning: a review of classification techniques. Informatika (2007)
24. Ortiz-Prado, E., Fernandez-Naranjo, R., Torres-Berru, Y., Lowe, R., Torres, I.: Exceptional prices of medical and other supplies during the COVID-19 pandemic in Ecuador. Am. J. Trop. Med. Hyg. **105**, 81–87 (2021). https://doi.org/10.4269/ajtmh.21-0221

HydroVLab, Hydrology Virtual Laboratory

Fernando Oñate-Valdivieso(✉), Santiago Quiñones, Jorge Fierro, and Fredy Cueva

Departamento de Ingeniería Civil, Universidad Técnica Particular de Loja, C/. Marcelino Champagnat S/N, 1101608 Loja, Ecuador
{fronate,lsquinones}@utpl.edu.ec

Abstract. Despite the Information and Communication Technology has been widely applied in online and distance learning, there are still topics, that involve the practical application of knowledge, that have not yet been fully resolved by ICTs, especially in the field of Engineering. This paper presents the Hydrology Virtual Laboratory (HydroVLab), an academic initiative that aims to provide tools for the simulation of hydrological phenomena, data analysis and the design of relevant elements in hydraulic and fluvial engineering through the Web. The didactic conception of HydroVLab allows students and researchers to analyze the different hydrological phenomena and their interrelation with their most relevant parameters. For the design, analysis and implementation of this software, the Unified Development Process was used, alongside a robust architecture that manages an assembly in three layers, which is developed under the advanced object-oriented language. This has allowed a flexible and adaptable design, capable of including new elements and performing changes for its expansion and maintenance.

Keywords: Hydro VLab · Virtual laboratory · Hydrology · Unified Development Process

1 Introduction

The 1960s witnessed the birth of the computer revolution and hydrological modeling. Computers provided the power to perform calculations that were not possible before, making possible the simulation of the entire hydrological cycle [1]. In the decades that followed, the evolvement of informatics made possible the development of easy-to-use software and tools for data acquisition, storage, retrieval, processing, and dissemination. Remote sensing tools emerged, such as radars and satellites, which made it possible to acquire spatial data for large areas [2–5]. In addition, geographic information systems (GIS) were developed to process large amounts of spatially distributed data [6]. The last two decades have witnessed the spread of artificial neural networks, fuzzy logic, genetic programming, and wave models [7–9].

The tools developed have always had an outlook towards research or engineering applications [1], being postponed the development of tools with an academic orientation, destined for professional training. In this context virtual laboratories play a preponderant role.

S. Berrezueta and K. Abad (Eds.): *Doctoral Symposium on Information and Communication Technologies - DSICT*, LNEE 846, pp. 128–140, 2022.
https://doi.org/10.1007/978-3-030-93718-8_11

A virtual laboratory can be defined as an electronic workspace conceived for remote collaboration and experimentation in order to research and carry out other creative activities, as well as elaborate and disseminate results through information and communication technologies [10]. Among the advantages of using virtual laboratories in the teaching-learning process, we can highlight: the methodological variety, the flexibility and easy access to computer applications, the attractive content presentation, the possibility of encountering new environments and problem statements, as well as the optimization of resources and costs [11]. Furthermore, a virtual laboratory makes it easier to carry out experimental activities with a greater number of students, without the need of their physical presence. The students then are capable of simulating many phenomena of different nature and modeling complex systems [12] such as hydrological cycle.

The Hydrology Virtual Laboratory (HydroVLab) integrates in a single framework the most common design, analysis and simulation processes in Hydrology, facilitating learning in a Web environment. The HydroVLab is based on a service-oriented architecture (SOA) [13], with a robust architecture distributed in three layers. All the different tools available in the Virtual Laboratory can be executed as an independent resource, maintaining a governance for the entire set of services.

2 Materials and Methods

2.1 Hydrologic Tools Implemented

The HydroVLab has three basic components:

a. A data analysis component, which includes a routine for missing information estimation through an orthogonal correlation analysis between stations. Prior gap filling, it is feasible to perform a graphical analysis of the climatic regime and station proximity, composing groups of stations and studying the homogeneity of their records using double-mass curves [14]. Additionally, tools are included for estimating extreme values by applying different probability distributions, selecting the optimal one for each case through goodness-of-fit tests [15].
b. A simulation component, which includes:

   - A tool to study the rainfall-runoff relationships based on the concept of the US SCS synthetic unit hydrograph [16, 17]. It is possible to observe the effect that some components have over the flows of a watershed, such as the magnitude of precipitation, the duration of the storm, as well as the soil type and use.
   - A tool to study the production of sediments through the Universal Soil Loss Equation [18–20] and sediment transport in natural channels considering both bedload and suspended sediment transport through the application of 24 different calculation methods [21].
   - A concentrated model of continuous simulation based on the Temez model [22, 23], ideal for basins with an area smaller than 300 km^2 [24].
   - A tool to calculate evapotranspiration using 8 methodologies and perform the water balance in the soil profile.

- A hydrologic event model using synthetic storm data or LAWR weather radar [4] data to determine flood flows.
- A tool for the simulation of hydrological flow routing in natural channels and reservoirs, applying commonly used methodologies such as Muskingum, Muskingum-Cunge [17], levelled out pool and kinematic wave [16].
- And finally, a tool to model data series using 5 different stochastic methods.

c. A design component, which has been implemented for the calculation of fundamental parameters in channel hydraulics such as the normal depth, critical depth, backwater curves and hydraulic jump, in rectangular, trapezoidal, and circular sections [25]. Additionally, a module has been implemented for the channelization and torrent control, applying river hydraulic criteria to calculate the regulated channel, the height of the maximum flood and the maximum scour depth in a channel subdivided into sections [26].

In addition, there is a Wiki that includes methodological details, fundamental concepts and theoretical contributions that facilitate the understanding of concepts and applications of HydroVLab. In order to encourage the exchange of opinions among users, individual discussion forums have been made available for each topic.

2.2 Computational Solution

The initial model for the design, analysis and simulation processes is presented in Fig. 1. The proposed and implemented model in the web solution can be seen in Fig. 2.

The processes that aimed to be automated were conceptualized and improved. Then, they were guided by a Unified Software Development Process [27] and its phases (Start, Elaboration, Construction and Transition) as shown in Fig. 3. The functional requirements of each tool were collected, the use cases were developed for each scenario the user can execute, and a flexible and adaptable architecture was designed to facilitate the implementation of future tools in HydroVLab.

A phase of review and improvement was carried out for each of the codified design and analysis methods in Visual Basic 6.0. This was done to implement and adapt the tools for its use through the Web. In a first prototype, the calculations were executed only from a local computer, nevertheless, an administration of input, processing and output files was implemented to optimize the storage resources and enable a Web environment.

Later, the interfaces were built for each virtual laboratory tool, which interact with the user in the following way:

- Entry of design parameters.
- Visualization of calculated data and generated graphs. This makes the environment intuitive and understandable.
- Execution of analysis methods that produce results that are displayed graphically and in plain text.
- Interpretation of results, conclusions, and recommendations.

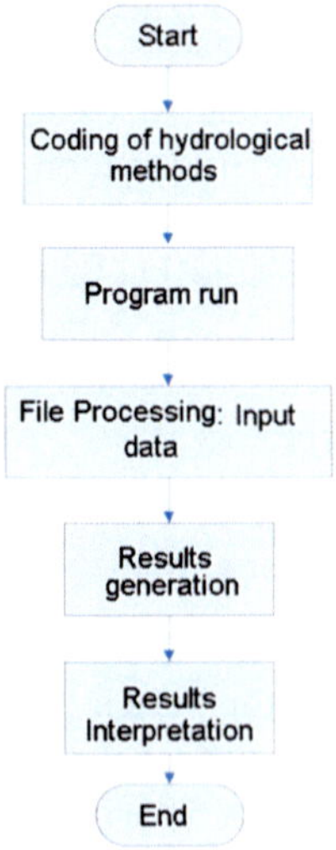

Fig. 1. Initial model for design, analysis and simulation.

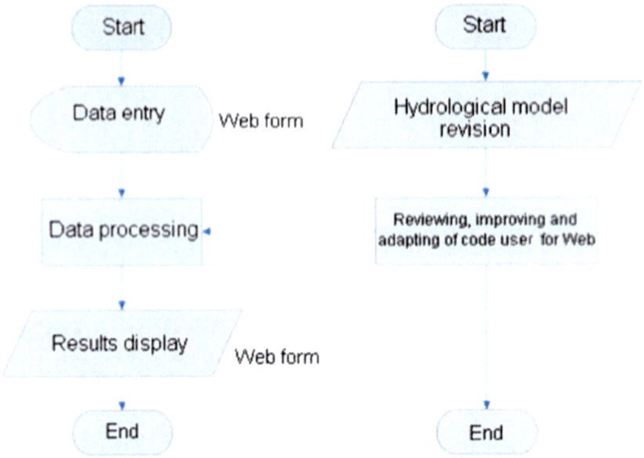

Fig. 2. Proposed model for hydrological design, analysis, and simulation.

Once the interfaces were developed, a Content Manager was implemented to manage the tools of the virtual laboratory, users, published information, among others. The Content Manager aimed to contain a set of collaborative tools focused on increasing the knowledge in hydrology and generally make the virtual solution an attractive, usable, and understandable portal for users.

All these requirements led to a stage of research and testing of the existing Content Administrative tools, concluding in the adoption of the open-source tool DotNetNuke 4.7 [28]. DotNetNuke is written in Visual Basic.Net and includes a public section (the portal itself) accessed by visitors, and a private section, accessed by page administrators, to edit the contents of the public section remotely.

Finally, the portal organization and configuration were defined as follows:

- General structure of the laboratory: User registration page, welcome and information page, data analysis, process simulation, design, collaboration forums, Wiki with updatable terminology and general information on authorship and the project in general.
- Each of the tools are publicly accessible, but their use requires prior registration.
- Once the users have finished their experimentation using any of the tools, a link is shown in which the results, conclusions and recommendations can be posted in a forum, which is organized by laboratory topic.

2.3 Architecture of the Computational Solution

A distributed architecture in "Three Layers" (Fig. 4) was adopted to develop the solution of the Hydrology Virtual Laboratory. Each layer is a set of components that fulfil a specific function. The decomposition of the solution aims to provide scalability, administration capacity, and better use of resources.

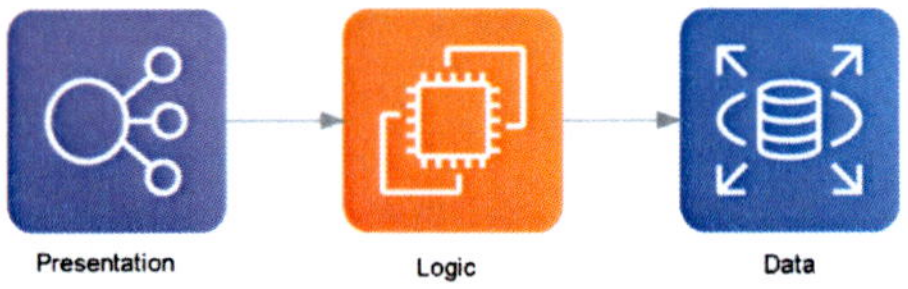

Fig. 3. Architecture of the solution.

For the development of the user interface, two options were analysed. The first one, using the DNN framework along with its logical and data layers that are embedded on the Website. And on the other hand, the development of a parallel and independent architecture, which was the preferred approach. Finally, our architecture was organized as follows: a.) "DotNetNuke" presentation layer and all its functionalities, b.) Logical layer for each HydroVLab tool and c.) Data access layer and functionalities developed for its use. The proposed architecture allows the incorporation of new tools easily and without major dependencies on the DNN website.

Figure 5 presents a global implementation view of the entire architecture of the solution, as well as the iteration and relationship of each of these layers with its predecessors and successors.

The development, implementation, and integration of each of the layers are described in detail in the following points.

Presentation Layer: The presentation layer comprises a series of pages for user interaction. In these pages the HydroVLab tools present information, accept inputs and display results. This user interface does not develop any processing or validation rules for the hydrological design and analysis, it relegates these processes to the business layer for manipulation.

It is worth mentioning that the integration of the DotNetNuke environment with the HydroVLab tools was developed independently of the website, that is, its own pages were created for each tool, using the DNN login module for the tools that require authorized access.

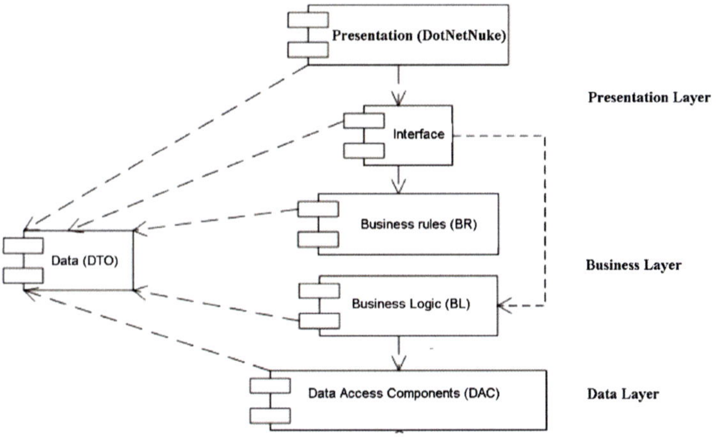

Fig. 4. Architecture distributed in three layers, for the hydrology virtual laboratory.

Business Layer: This layer applies its own business rules for hydrological design, analysis, and simulation processes. Among them we have:

- Execution of hydrological analysis methodologies.
- Calculation of extra design parameters necessary for the analysis.

Data Layer: In this layer we contemplate the Data Access Logic for the Content Manager of the Hydrology Virtual Laboratory.

It contains all requests to the database, allowing the DotNetNuke portal to communicate with the data source through the Data Provider, the concrete data providers, and the application blocks. The Data Provider is an abstract class, and the concrete data providers inherit the Data Provider class to perform operations with the database. Concrete Data Providers allow to write Data Providers for any database such as Oracle, Microsoft SQL Server, MySQL, etc. The data access application block simplifies the development tasks that implement the common data access functionality [28].

2.4 Development of the Computational Solution

For the development of the computational solution, several methodological tools and computer techniques were used for the design, analysis, development, and implementation of HydroVLab in each of its stages. The application and use of these tools and methodologies are described below.

The Unified Software Development Process (Fig. 3) is a development methodology guided by time, human resources, project size, among other factors, that was effectively adapted to develop this solution.

The HydroVLab tools were designed for its implementation based on object-oriented programming and the Unified Modelling language. The Unified process enables the configuration and adaptation of the project to the size, complexity and changes that have changed along its development.

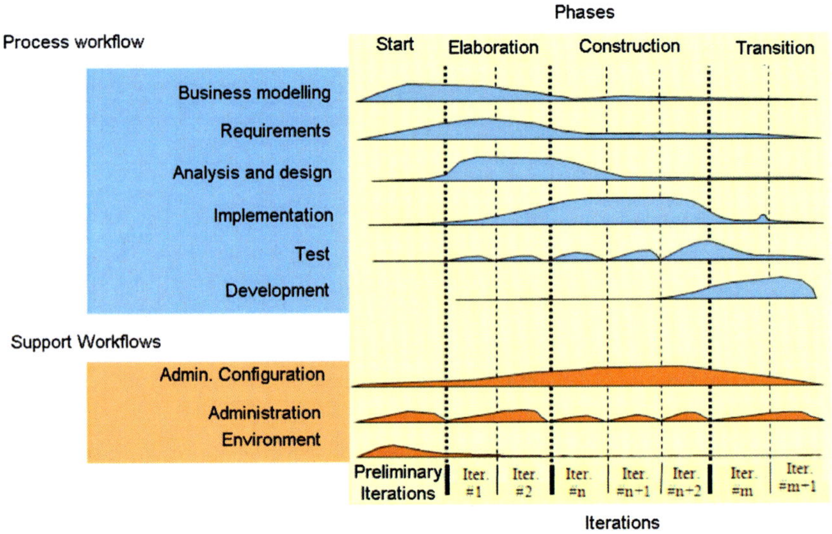

Fig. 5. Unified software development process (PUDS)

The Unified Modelling Language (UML) [29] made it possible to visualize the structure and behaviour of the HydroVLab architecture and its components, detailed in the development phase to build each of the following elements:

- Use cases: To model the hydrological design, analysis, and simulation tools.
- Sequence diagrams: To model the messages transferring between objects.
- Class diagrams: To model the static structure of the classes in the solution.
- Physical/Implementation diagrams: To model the distribution of the solution.
- The database used for the virtual laboratory was Microsoft SQL, due to its high performance, reliability, security, and scalability. Among the advantages of using this technology for the laboratory we can highlight:
- Its language and region versatility make the virtual laboratory a global platform.
- Being a robust database, its architecture guarantees the reliability and security of the data.
- Its security and access schemes through privileges enables a safe management of the database objects (tables, views, etc.), users and roles assigned for the use of each tool in the virtual laboratory.

The design and analysis methods were coded in Visual Basic.Net. Web applications use ASP.NET from Visual Studio 2008, 2015. The logical and data access layers were also coded in Visual Basic.Net. The advantages of using this programming language for the virtual laboratory are, among others:

- The tools are developed with a powerful object-oriented language.
- The incorporation of a variety of objects allows the interconnection with different DB engines.
- It is possible to separate code from graphical interface and business logic.

- The tools are implemented on a multilayer distributed architecture.
- Each tool developed is compatible with DotNetNuke 4.7.
- It allows a 40% use of JavaScript, which makes it possible to execute processes on the client, making the response time of each tool minimal.
- It allows the use of AJAX, which merges XML and JavaScript allowing the simulation of a Windows application, avoiding "the blank screen" while the process is being processed on the server.

2.5 Validation of HydroVLab's Tools

The portal and tools have been exposed to functionality tests and graphical integrity tests, with special attention to DotNetNuke portal.

The numerical efficiency of the calculations performed by the HydroVLab tools was validated by direct comparison with the results obtained applying similar operating software (HEC-HMS, HEC-RAS, CHAC, CropWat, Sams, among others) and electronic sheets. The HydroVLab intuitive usage was evaluated by 120 students from the School of Civil Engineering of the Universidad Técnica Particular de Loja and their opinions were collected through anonymous surveys.

3 Results

Once the process of design, analysis, development, implementation, and testing of the Hydrology Virtual Laboratory was completed, it was possible to implement a Bilingual Web Virtual Laboratory (Spanish - English) with tools for data analysis, process simulation, and design of works related to hydrology and river engineering (Fig. 6). The HydroVLab can be found at http://www.hydrovlab.utpl.edu.ec.

As an example, Fig. 7 shows a window of the simulation component of HydroVLab, for the study of rainfall-runoff relationships. This tool is used to see the effect of the effective precipitation's magnitude on flood flows in a basin. HydroVLab allows to vary the initial conditions and to observe graphically and numerically the effect of these changes. In this way, the sensitivity that each of the hydrological processes present to variations in their boundary conditions can be studied.

Figure 8 shows the hydrological flow routing module using the Muskingum method, making it possible to calibrate the value of the K and X coefficients as a function of observed hydrographs. Once the value of the coefficients is known, the flow routing is carried out. It should be noted that the calibration of the parameters and the hydrological flow routing is interactive, allowing the input data and the value of the parameters to be changed at any time, enabling the visualisation of their interrelationships.

Among the design tools available in HydroVLab, the tool for dimensioning of the regulated channel for channelization and torrent control stands out, allowing the horizontal and vertical design of the civil works required for this purpose. Data entry is done using text files that include the geometric characteristics of each cross section and its Manning roughness coefficient. Graphically it presents the longitudinal profile of the section, the free surface depth, the regulated channel, and the detail of the necessary transversal works. Part of the process can be seen in Fig. 9.

Fig. 6. Hydrology virtual laboratory - welcome page.

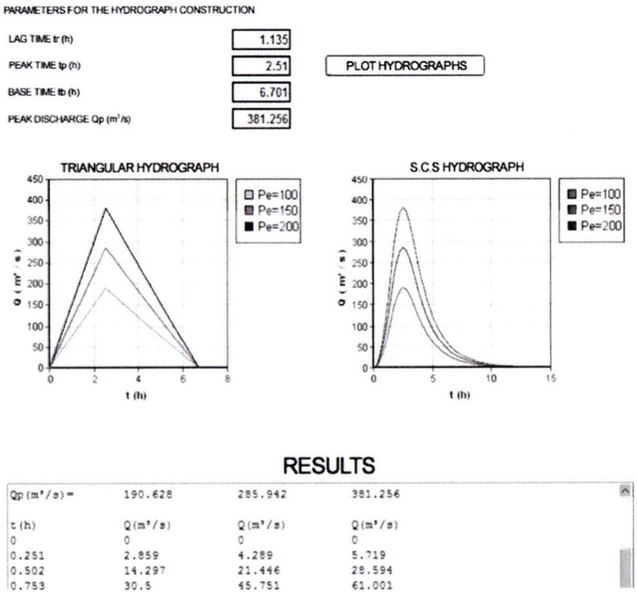

Fig. 7. Generation of flood hydrographs in HydroVLab.

The tool calculates the maximum flood levels and determine the size of the regulated channel section. Finally, it determines the location and dimensioning of the transversal works for the vertical control of the slope.

Figure 10 shows the data analysis tool for the homogenization of historical series.

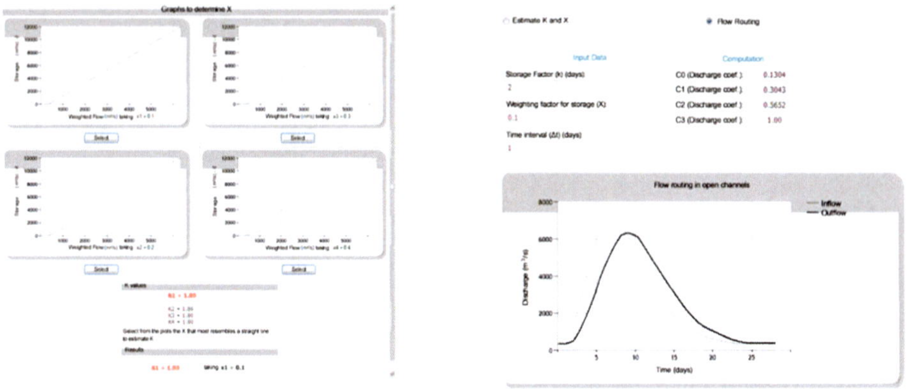

Fig. 8. Flow routing using the Muskingum method in HydroVLab.

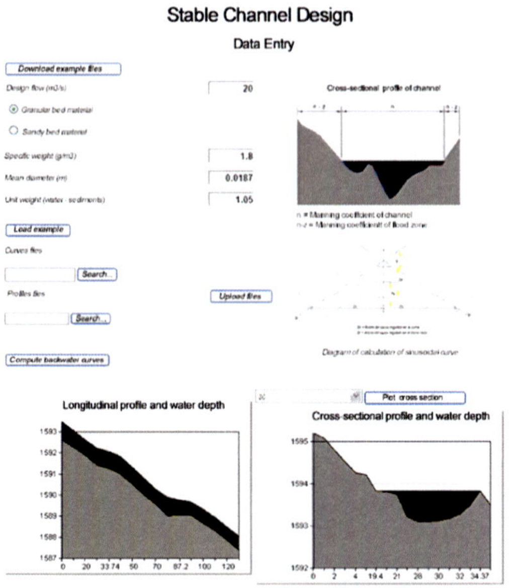

Fig. 9. Design of the regulated channel in HydroVLab.

This tool has provided: a.) The graphic representation of the seasonal variation of the records. b.) The comparative representation of the extension of the data, which allows to identify common periods of information c.) The spatial location of the stations to be used in the analysis, d.) The interactive construction of double mass curves and e.) The homogenization of historical series by an orthogonal correlation analysis. This tool requires a special text file format for data entry since it allows the simultaneous handling of the historical record of up to 20 stations.

Regarding validation, among the most important functionality tests conducted for the DotNetNuke portal, we can highlight that of user administration management, in which the correct assignment of roles and credentials for access to the tools is validated.On

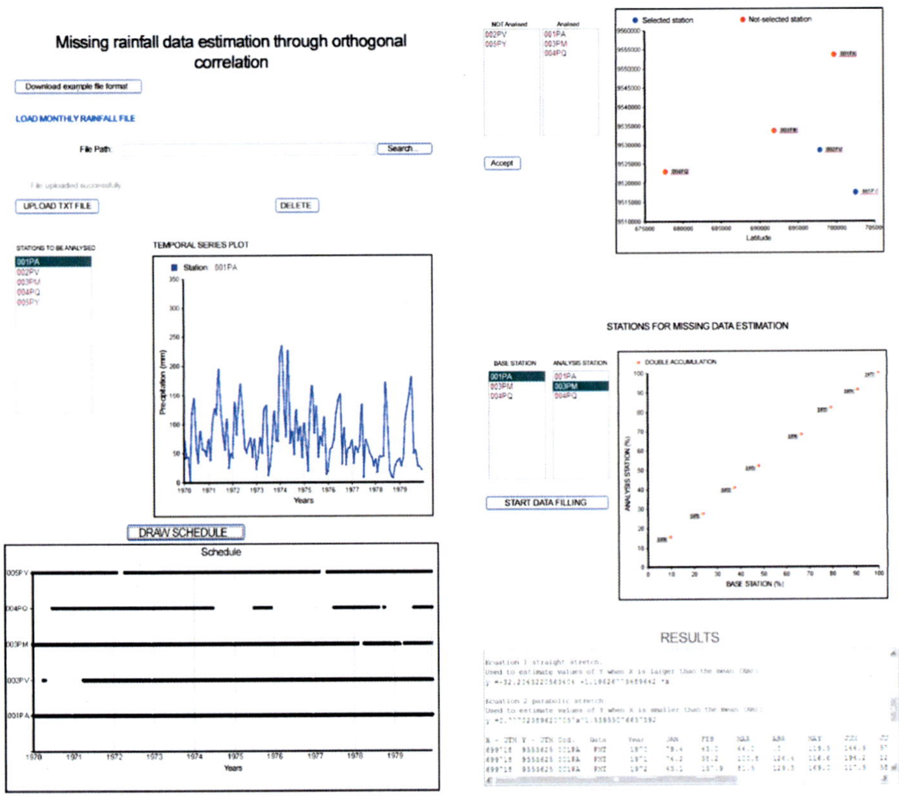

Fig. 10. Data analysis for homogenization of records in the HydroVLab.

the other hand, the graphical integrity tests validate the compatibility of tools and portal with the most used browsers (Chrome, Firefox, Microsoft Edge) with load times of less than 5 s.

The comparison of graphical and numerical results obtained through HydroVLab and those obtained with software with similar characteristics yielded 100% correspondence, observing in some cases minimal differences attributed to small methodological variations introduced in the software used for comparison. Regarding the intuitive usage of the environment and its tools, around 80% of the evaluators gave it a high rating, about 15% gave it a medium rating and 5% a low rating. These results support the design of the virtual laboratory, considering that it is sufficiently friendly, and could be used in distance learning without major inconveniences.

4 Conclusions

HydroVLab enables the study of hydrological phenomena and the practical application of the knowledge acquired in the academic field, breaking geographical and temporal barriers; creating a space that enriches distance learning with on-line support of topics related to Hydrology and Water Resources, constituting a differentiating element in e-learning.

With the proposed computational solution, it has been possible to integrate the hydrological design, analysis, and simulation processes into a single framework on a Web environment that can be accessed from anywhere with internet connection, regardless of the operating system used. The architecture managed and implemented during and after the laboratory development offers the possibility of integrating further hydrological methods, as well as their respective graphical environments. The use of DotNetNuke allowed the creation of websites with features and characteristics that ensure an easy interaction between the user and the services that are implemented on the site.

References

1. Singh, V.P.: Hydrologic modeling: progress and future directions. Geosci. Lett. **5**(1), 1–18 (2018). https://doi.org/10.1186/s40562-018-0113-z
2. Hogg Y., Zhang Y., Khan S.I.: Hydrologic remote sensing: capacity building for sustain
3. ability and resilience. CRC Press, Boca Raton (2017)
4. Lakshmi, V.: Remote Sensing of Hydrological Extremes. Springer, Berlin (2017)
5. Oñate-Valdivieso, F., et al.: Temporal and spatial analysis of precipitation patterns in an Andean region of southern Ecuador using LAWR weather radar. Meteorol. Atmos. Phys. **130**(4), 473–484 (2017). https://doi.org/10.1007/s00703-017-0535-8
6. Mejía-Veintimilla, D., et al.: River discharge simulation in the high Andes of southern Ecuador using high-resolution radar observations and meteorological station data. Remote Sens. **11**(23), 2804 (2019)
7. Maidment, D.R.: Arc Hydro. ESRI Press, Redlands (2002)
8. Kumar, P., Alameda, J.C., Bajcsy, P., Folk, M., Markus, M.: Hydroinformatics: Data Integrative Approaches in Computation, Analysis, and Modelling. Taylor & Francis, Boca Raton (2006)
9. Ross, T.J.: Fuzzy Logic with Engineering Applications. John Wiley, New York (2010)
10. Tayfur, G.: Soft Computing in Water Resources Engineering: Artificial Neural Network, Fuzzy Logic and Genetic Algorithms. WIT Press, Southampton (2012)
11. Vary, J.P.: Report of the Expert Meeting on Virtual Laboratories, organized by the International Institute of Theoretical and Applied Physics @TAP), Ames. Iowa. UNESCO, Paris (2000)
12. Infante Jiménez, C.: Propuesta pedagógica para el uso de laboratorios virtuales como actividad complementaria en las asignaturas teórico-prácticas. Rev. Mex. Investig. Educ. **19**(62), 917–937 (2014)
13. Lorandi, A., Hermida, G., Hernández, J., Ladrón de Guevara, E.: Los laboratorios virtuales y laboratorios remotos en la enseñanza de la ingeniería. Revista Internacional de Educación en Ingeniería **4**, 24–30 (2011)
14. Contreras Castro, M.D.: Prácticas académicas de la arquitectura orientada al servicio (SOA). Investigación. e Innovación en Ingeniería **7**(1), 60–97 (2019)
15. Dingman, S.L.: Physical Hydrology, 2da edición. Prentice Hall, New Jersey (2002)
16. Coronado-Hernández, O.E., Merlano-Sabalza, E., Díaz-Vergara, Z., Coronado-Hernández, J.R.: Selection of hydrological probability distributions for extreme rainfall events in the regions of Colombia. Water **12**(5), 1397 (2020)
17. Chow, V.T., Maidment, D., Mays, L.: Hidrología Aplicada. McGraw-Hill, Bogotá (1994)
18. Ponce, V.M.: Engineering Hydrology. Prentice Hall, Englewood Cliffs, NJ (1989)
19. Renard, K.G., Foster, G.R., Weesies, G.A., y Porter, J.P.: Revised universal soil loss equation. J. Soil. Water Conserv. **46**, 30–33 (1991)
20. Wischmeier, W.H., Smith, D.D.: Predicting Rainfall Erosion Losses. Agriculture Handbook 537. United States Department of Agriculture. Science and Education Administration (1978)

21. Oñate-Valdivieso, F., Sendra, J.B.: Semidistributed hydrological model with scarce information: application to a large south american binational basin. J. Hydrol. Eng. **19**(5), 1006–1014 (2014)
22. Maza, J., García, M.: Transporte de Sedimentos. Instituto de Ingeniería UNAM, México (1996)
23. Témez, J.R.: Modelo matemático de Transformación Precipitación Aportación. Asociación de Investigación Industrial Eléctrica ASINEL, Madrid (1977)
24. Estrela Monreal, T. (1999). "Los modelos de simulación integral de cuenca y su utilización en estudios de recursos hídricos". Ingeniería Civil. 72, p. 83–95. Madrid
25. Oñate-Valdivieso, F., Bosque-Sendra, J., Sastre-Merlin, A., Ponce, V.M.: Calibration, validation and evaluation of a lumped hydrologic model in a mountain area in Southern Ecuador. Agrociencia **50**(8), 945–963 (2016)
26. Chow, V.T.: Hidráulica de los Canales Abiertos. McGraw-Hill, Bogotá (1994)
27. Martín Vide, J.P.: Ingeniería Fluvial. Escuela Colombiana de Ingeniería, Bogotá (2000)
28. Shuja. A.K., Krebs, J.: IBM Rational Unified Process Reference and Certification Guide: Solution Designer. IBM Press/Pearson (2008)
29. DotNetNuke, Progress with DotNetNuke 4.7.0 > DNN Corp. https://www.dnnsoftware.com/community-blog/cid/135578/progress-with-dotnetnuke-470, Accessed 21 June 2021
30. La guía sencilla para la diagramación de UML y el modelado de la base de datos. https://www.microsoft.com/es-ww/microsoft-365/business-insights-ideas/resources/guide-to-uml-diagramming-and-database-modeling, Accessed 21 June 2021

Author Index

S. Berrezueta and K. Abad (Eds.): *Doctoral Symposium on Information and Communication Technologies - DSICT*, LNEE 846, p. 141, 2022.
https://doi.org/10.1007/978-3-030-93718-8